NUCLEUS
ENGLISH FOR SCIENCE AND TECHNOLOGY
GENERAL SCIENCE

The other titles in this series are:

Biology	Donald Adamson/Martin Bates
Geology	Colin Barron/Ian Stewart
Engineering	Tony Dudley-Evans/Tim Smart/John Wall
Agriculture	Stephen Denny/Lewis Kerr/ Martin Phillips/ Clarence Shettlesworth
Chemistry	Colin Barron/Ian Stewart
Physics	Tim Bowyer/David Hall
Mathematics	Tim Bowyer/David Hall
Medicine	Jeffrey Jameson/David Kirwan/Tony O'Brien
Nursing Science	Rosalie Kerr/Jennifer Smith

The above authors were all members of the team who worked together with the Editors on the Series. The Editors would like to thank them, together with Dick and Margaret Knight, for certain suggestions incorporated into *General Science*.

NUCLEUS

ENGLISH FOR SCIENCE AND TECHNOLOGY
GENERAL SCIENCE

Martin Bates
Tony Dudley-Evans

Series Editors
Martin Bates and Tony Dudley-Evans
Science Adviser to the Series
Arthur Godman C. Chem., MRIC

Longman

Longman Group Limited
London

*Associated companies, branches and
representatives throughout the world*

© Longman Group Ltd 1976

All rights reserved. No part of this publication may be reproduced, stored in a retrieval system, or transmitted in any form or by any means, electronic, mechanical, photocopying, recording, or otherwise, without the prior permission of the Copyright owner.

First published 1976
4th impression 1978

ISBN 0 582 51300 6

Printed in Great Britain
by J. W. Arrowsmith Ltd.,
Bristol

Contents

		page
Unit 1	Properties and Shapes	1
Unit 2	Location	11
Unit 3	Structure	18
Unit A	Revision	26
Unit 4	Measurement 1	28
Unit 5	Process 1 Function and Ability	36
Unit 6	Process 2 Actions in Sequence	44
Unit B	Revision	54
Unit 7	Measurement 2 Quantity	57
Unit 8	Process 3 Cause and Effect	67
Unit 9	Measurement 3 Proportion	78
Unit C	Revision	86
Unit 10	Measurement 4 Frequency, Tendency, Probability	89
Unit 11	Process 4 Method	98
Unit 12	Consolidation	107
Glossary		117

Unit 1 Properties and Shapes

Section 1 One-dimensional and two-dimensional properties

1. Look at these:

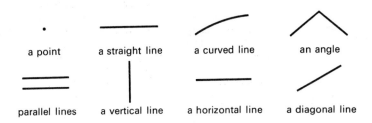

Now read this and answer the questions:

> The letter 'E' has one vertical line and three horizontal lines. It also has four angles.
> Which of these letters are described below?
> D, M, C, H, F, L, Z, B.

a) A letter with 2 horizontal lines and 1 vertical line.
b) A letter with 1 curved line and no straight lines.
c) A letter with 2 curved lines and 1 vertical line.
d) A letter with 2 parallel vertical lines, 1 horizontal line and 4 angles.
e) A letter with 2 vertical lines and 2 diagonal lines.

Now write sentences describing these signs:

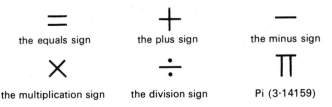

2. Look at these figures and answer the questions:

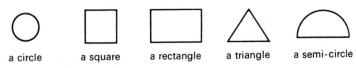

a) Which figure is curved?
b) Which figures have parallel sides?
c) Which figure always has equal sides?

d) **Which figure may have equal sides?**
e) **Which figure has 3 angles?**
f) **Which figure has a curved side and a straight side?**

Now make sentences from the table:

Example: A coin is shaped like a circle. It is circular in shape.

A coin		square.		rectangular	
A ruler		rectangle.		circular	
A set square	is shaped like a	semi-circle.	It is	square	in shape.
A protractor		triangle.		semi-circular	
A chess-board		circle.		triangular	

3. Look at this plan of a town:

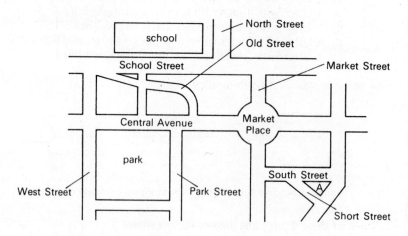

Answer these questions:

a) What shape is the plan of the school?
b) Which street is curved?
c) What shape is area A?
d) Which area is square?
e) Name two streets which are parallel.

f) Are Old Street and School Street parallel?
g) Which part is roughly circular in shape?
h) Which streets meet at an angle of 90 degrees (at right angles)?
i) Which streets meet at a different angle?

Section 2 Three-dimensional shapes

4. Look and answer:

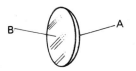

This is a lens. One *surface* is *curved* and the other is *flat*. Which is which?

Look at these solids:

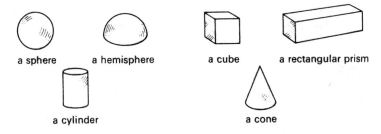

Now describe them:

Example: A cube has 6 surfaces. They are all flat and square.

5. Look and read:

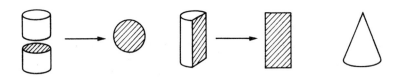

The *cross-section* of a cylinder is circular. The *longitudinal* section is rectangular. The sides of a cylinder are parallel. The sides of a cone are *tapering*.

Answer these questions:

a) What shape is the cross-section of a sphere?
b) What shape is the longitudinal section of a hemisphere?
c) What shape is the cross-section of a cube?
d) Which solid is rectangular in cross-section?
e) In longitudinal section, are the sides of a cylinder parallel or tapering?

f) In longitudinal section, are the sides of a cone parallel or tapering?
g) What shape is the cross-section of a cone?

6. Complete these:

    Cylindrical = shaped like a _____
    Cubic   = shaped like a _____
    Conical   = shaped like a _____
    Spherical   = shaped like a _____

Now describe the shapes of these objects:

Example: A ball is spherical in shape.

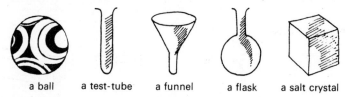

a ball   a test-tube   a funnel   a flask   a salt crystal

7. Look at this:

This tube is shaped like the letter 'U'.
It is *U-shaped*.

Describe the shapes of the following:

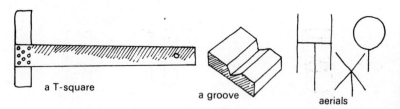

a T-square   a groove   aerials

These objects are used to describe shapes:

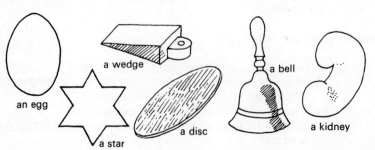

an egg   a wedge   a bell   a kidney
a star   a disc

Now describe the following objects:

Example: A potato is egg-shaped.

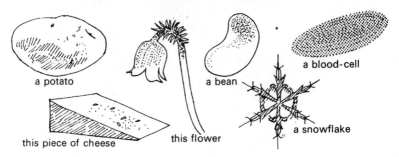

8. Look at this picture:

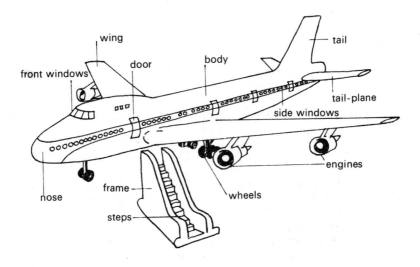

Now say whether these statements are true or false. Correct the false statements.

    a) The tail is nearly triangular in shape.
    b) The door is flat.
    c) The steps are parallel to each other.
    d) The sides of the frame are curved.
    e) The tail-plane is wing-shaped.
    f) All the windows are circular.
    g) The engines are nearly cylindrical.
    h) The wheels are cubic in shape.
    i) The front of the plane is cylindrical.
    j) The nose is tapering.
    k) The wings are at right angles to the body.

9. Look at these diagrams and complete the descriptions:

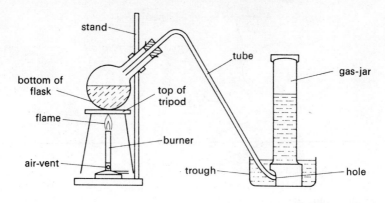

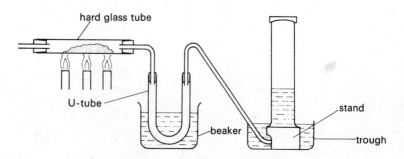

In the first apparatus, the bottom of the flask is _____ in shape. The flask is in a _____ position. The stand is _____. The gas-jar is _____ in shape. The burner is also _____. The air-vent is _____. The flame is _____. The bottom of the trough is _____. The hole at the bottom of the gas-jar is _____ in shape. The top of the tripod is _____. In cross-section, the tube is _____.

Now make as many sentences as you can describing the second apparatus.

Section 3 Properties of materials

10. Look and read:

Ice is *solid*. Water is *liquid*. Steam is *gaseous*. Steam and water are *fluids*.

```
                  ┌─────────┐
          −183°C  │ Oxygen  │  Boiling Point
        −218·4°C  │ Oxygen  │  Melting Point
                  │ Oxygen  │
                  └─────────┘

                  ┌─────────┐
                  │  Neon   │
        −245·9°C  │  Neon   │  Boiling Point
        −248·7°C  │  Neon   │  Melting Point
                  └─────────┘
```

Complete these statements:

    a) At −183°C oxygen changes from the gaseous state to the _____ state.
    b) At −218·4°C oxygen changes from the liquid state to . . . .
    c) At 183°C oxygen is in the _____ state.
    d) At −246°C neon is in the _____ state.
    e) At −220°C oxygen is in the _____ state.
    f) Steam, water, ice, oxygen, neon: all these are fluids except _____.

11. Read the following properties of materials and complete the examples:

A *brittle* material *breaks* easily; eg glass, . . .

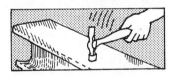
A *tough* material does not break easily; eg steel, . . .

A *hard* material is difficult to *scratch*; eg glass, . . .

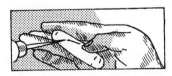
A *soft* material is easy to scratch; eg chalk, . . .

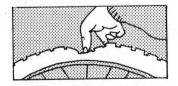
A *flexible* material *bends* easily; eg rubber, . . .

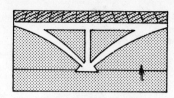
A *rigid* material does not bend easily; eg concrete, ...

Answer these questions:

a) Why does a glass beaker break if you drop it?
b) Why doesn't a polythene beaker break?
c) Why is butter easy to cut?
d) Why can a diamond cut glass?
e) Why do the branches of a tree bend in the wind?
f) Why don't the walls of a house bend in the wind?
g) Which is more flexible: a wooden ruler or a plastic ruler?
h) What are the different properties of green wood (on a tree) and dry wood?

12. Read and complete these:

Some materials have a *smooth* surface; they produce little *friction* when they are rubbed; eg ice, ...

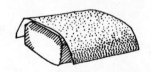
Some materials have a *rough* surface and produce a lot of friction; eg sandpaper, ...

Materials which are *soluble* in water *dissolve* easily; eg salt, ...

Materials which are *insoluble* do not dissolve; eg glass, ...

You can see through *transparent* materials; eg water, . . .

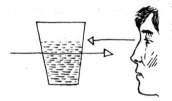
You cannot see through *translucent* materials but the light passes through them; eg dirty water, . . .

You cannot see through *opaque* materials and the light cannot pass through them; eg metal, . . .

Combustible materials *burn* easily; eg wood, . . .

Non-combustible materials do not burn, eg stone, . . .

Read this and choose the right properties:

    A material which is used for making clothes must be solid/fluid, brittle/tough, soft/hard, rigid/flexible, smooth/rough, opaque/transparent and soluble/insoluble.

Complete these sentences:

    One material with these properties is wool. Others are _____ and _____.

    Steel is not generally used for clothes because it is . . . .
    Glass is unsuitable because it is _____, _____ and _____.

Now suggest different properties which are suitable for the following purposes and give examples of materials with the right properties:

a) For the body of a car we need a material which is _____, _____, _____ and _____, eg _____.
b) For a window ... , eg _____.
c) For a cooking pot ... , eg _____.

13. Complete the following table, giving the properties of the materials:

	steel	glass	rubber	sugar	wood
tough/brittle	tough				
soft/hard	hard				
soluble/insoluble	insoluble				
combustible/non-combustible	non combustible				
flexible/rigid	rigid				
transparent/opaque	opaque				

Look at these examples and make other questions and answers like them:

Example: What properties have glass and steel in common?
Glass and steel are hard, insoluble and rigid.

Unit 2 Location

Section 1 Positions on two dimensions

1. Look and read:

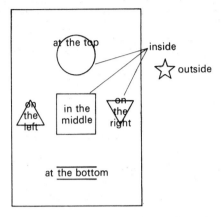

The words give the positions of the shapes *in relation to the rectangle*.

Make questions and answers like the following:

Example: What is there *at the top of* the rectangle?
There is a circle at the top of the rectangle.

2. Now look at this:

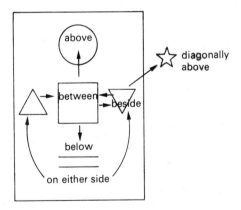

The words give the positions of the shapes *in relation to one another*.

Make questions and answers like the following:

Example: Where is the circle?
The circle is above the square.

3. Look and read:

Sc Scandium	Ti Titanium	V Vanadium	Cr Chromium	Mn Manganese	Fe Iron	Co Cobalt	Ni Nickel	Cu Copper	Zn Zinc
Y Yttrium	Zr Zirconium	Nb Niobium	Mo Molybdenum	Tc Technetium	Ru Ruthenium	Rh Rhodium	Pd Palladium	Ag Silver	Cd Cadmium
La Lanthanum	Hf Hafnium	Ta Tantalum	W Tungsten	Re Rhenium	Os Osmium	Ir Iridium	Pt Platinum	Au Gold	Hg Mercury

Above there is a table of some elements. The elements are arranged in horizontal *rows* and vertical *columns*.

Give the positions of the following elements in relation to the whole table:

Examples: Lanthanum is at the bottom, on the left.
Vanadium is *in the third column from the left*, at the top.
Cobalt is *in the top row, near* the middle.

Tungsten, cadmium, zinc, gold, scandium, iron.

Now give the position of these elements in relation to others:

Example: Osmium is beside and *to* the right of rhenium.

Cobalt in relation to nickel and iron
Niobium in relation to molybdenum
Platinum and mercury in relation to gold
Gold in relation to silver
Iron in relation to rhodium
Silver in relation to zinc
Silver in relation to gold

4. Read these sentences which give other positions:

Cobalt is *next to*, or *adjacent to*, nickel.
Iron is not adjacent to nickel because cobalt is between them.
Manganese is *in line with* copper and gold is in line with hafnium.
Yttrium is *near* tantalum but *far from* zinc.

Now say whether these statements are true or false. Correct the false statements.

a) Silver is diagonally above nickel.
b) Zinc is in line with scandium.
c) Molybdenum and ruthenium are on either side of technetium.
d) Gold is adjacent to mercury.
e) Iron is beside and to the right of cobalt.
f) Gold is vertically below silver.

g) Vanadium is near cadmium.
h) Mercury is at the bottom of the table, on the right.
i) Copper is between nickel and zinc.
j) Manganese is in the middle row.
k) Silver is in the third column from the right.

Section 2   Positions on three dimensions

5. Look and read:

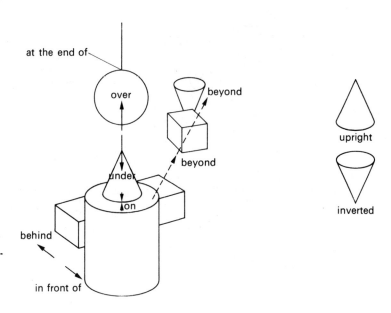

Note: The sphere is *above* all the other solids, but it is *over* only the cone and the middle of the cylinder.

Now make questions and answers:

Example: Where is the sphere in relation to the upright cone?
The sphere is *over* the upright cone.

sphere ... upright cone.   upright cone ... cylinder.
upright cone ... sphere.   cylinder ... upright cone.
rectangular solid ... cylinder.   inverted cone ... cube.
cylinder ... rectangular solid.   cube ... inverted cone.
cube ... rectangular solid.   sphere ... line.

6. Look and read:

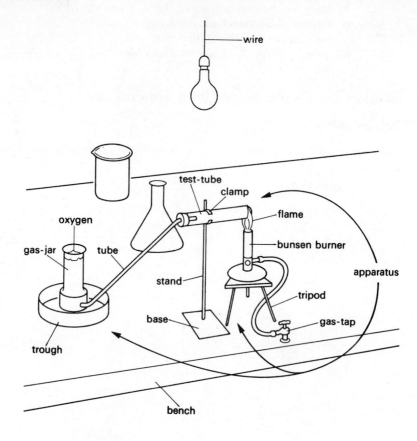

Now complete this description:

a) There is a gas-jar ... of the apparatus.
b) The gas-jar is standing _____ a trough.
c) There is some oxygen _____ the gas-jar.
d) ... of the gas-jar there is a stand.
e) ... of the stand there are two clamps which hold a test-tube.
f) _____ the test-tube and the gas-jar there is a tube.
g) ... the stand there is a base.
h) The base is _____ the test-tube.
i) There is a tripod and a bunsen burner ... the stand.
j) The burner is _____ the tripod.
k) The gas comes from a tap which is ... of the tripod.
l) ... the burner there is a flame.
m) _____ the apparatus there is a conical flask.
n) _____ the conical flask there is a beaker.
o) There is a light _____ the stand.
p) The light is ... of a wire.
q) The light is _____ the apparatus and the flasks.
r) The apparatus and the flasks are _____ the bench.

Section 3 Geographical positions

7. Look at this map of the world and the sentences below it:

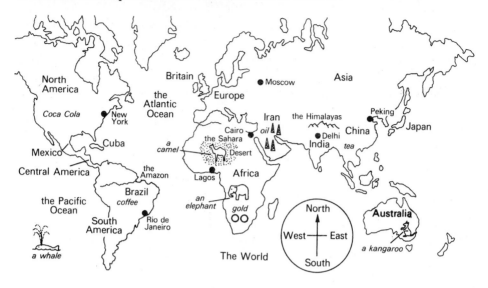

Mexico *is situated in* Central America
Central America is situated *between* North America and South America.
Europe is situated *to the west of* Asia.
The Sahara Desert *is located* in Africa and *to the south-west of* Cairo.
Gold *is found in the south of* Africa.
Kangaroos are found in Australia.
People *are distributed throughout* the world.

Answer these questions:

    a) Where are whales found?
    b) Where is the River Amazon located?
    c) Where is Lagos situated?
    d) Where is the Atlantic Ocean in relation to Europe and North America.
    e) Where are the Himalayas located in relation to China?
    f) Where is tea found?
    g) Where is Moscow situated in relation to Delhi?
    h) Where is India situated in relation to Asia?

Ask and answer some more questions like these.

Write six sentences about your own country, using these words:

    is/are situated   is/are distributed throughout
    is/are located   to the east of
    is/are found   in the north of

Section 4 Some parts of objects and their properties

8. Look and read:

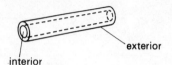

The *interior* of a glass tube is *hollow*.
The *exterior* is hard and smooth.

Ask and answer questions with the words below:

Example: Which part of an egg is hard?
The exterior of an egg is hard.

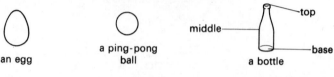

an egg	a ping-pong ball	a bottle
hard/soft	hollow/tough	hard and transparent/ flat/circular/cylindrical/ hollow

Look and read:

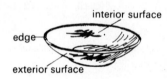

The interior surface of this dish is *concave*.
The exterior surface is *convex*.
The *edge* is circular.

This book is *thick*.
The pages are *thin*.
The edges of the pages are straight.

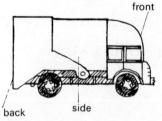

The *front* of this lorry is curved.
The *back* and the *sides* are flat.

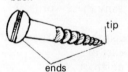

One *end* of this screw is flat and circular.
The other end is tapering.
The *tip* is *pointed*.

Now ask and answer more questions:

a convex lens   a concave lens   a leaf   a pencil

flat/convex/   concave/   curved/   filled with graphite/
thick/thin   thick/thin   tapering smooth/tapering/
                                             pointed/flat

Look at the diagrams and complete the descriptions:

The _____ and the _____ of this case are horizontal.
The _____, the _____ and the _____ are vertical.
There is a handle on the _____.
The _____ is curved.
The _____ is hollow.

The _____ of a cigarette is filled with tobacco.
The _____ are circular.
The _____ is white.

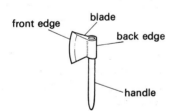

The _____ edge of the blade is thin.
The _____ edge is thick.
The blade is _____ -shaped.
The _____ of the handle is tapering.

Unit 3 Structure

Section 1 Parts and the whole

1. **Look and read:**

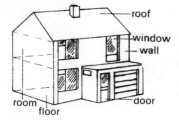

A house *consists of* walls, a roof, floors, doors and windows. (These are the *parts* of the house.)
It *contains* rooms. (The rooms are inside the house.)

Now complete this:

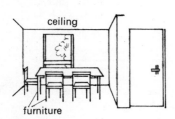

A room _____ walls, a ceiling, a floor, a _____ and _____.
A room often _____ furniture.

Answer these:

What does your classroom consist of?
What does it contain?

Complete this:

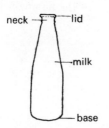

A milk-bottle consists of a glass cylinder, a flat _____, a tapering _____ and a lid.
It contains _____.

Answer these:

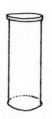

What does a gas-jar consist of?
What does it contain?

Complete this:

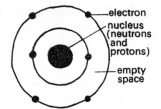

An atom of carbon consists of . . . .
It contains a _____ in the centre.
The nucleus consists of _____ and _____.

2. Read this:

The rooms in a house *include* a bedroom, a sitting-room *etc.* (These are some of the different kinds of room.)

Complete these:

The rooms in a school include . . .

Furniture includes . . .

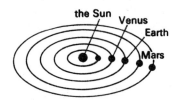

The solar system _____ the Sun and planets. Planets _____ the Earth, Mars, Venus _____.

Look and complete:

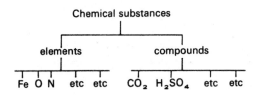

Chemical substances consist of _____ and _____.
Elements include . . . .
Compounds include . . . .

Now read the text and copy out the complete diagram:

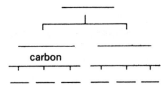

Matter consists of organic substances and inorganic substances. Organic substances include coal and oil. Inorganic substances include iron and sulphur. Organic substances contain carbon. Inorganic substances do not contain carbon.

3. Say whether these statements are true or false. Correct the false statements.

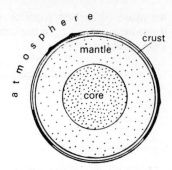

a) The Earth consists of a core and an atmosphere.
b) The crust is part of the Earth's structure.
c) Other parts of the Earth include the mantle, the core and the Sun.
d) The core contains the Earth.
e) The atmosphere contains gases.

Section 2 The connection between parts

4. Look and read:

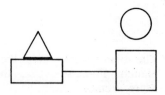

The rectangle *is connected to* the square *by* the line.
The triangle *is attached to* the rectangle.
The circle *is detached from* the square.

Now complete these:

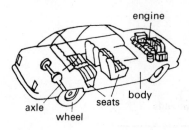

A car _____ a body, seats, an engine, wheels, axles.
The front of this car _____ the engine.
The wheels are _____ by the axles.
The wheels are _____ to the axles.
The wheels are _____ from the body.
Different kinds of car _____ Mercedes, Moskvich etc.

Look and read:

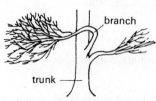

The branch of the tree *is joined to* the trunk.

The branches *are supported by* the trunk.

20

Now complete this description:

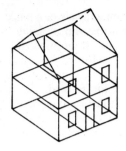

This house consists of . . .
It contains
The rooms include . . . .
The roof is supported by . . . .
The walls are joined to . . . .
This house is shaped like . . . .
The windows are situated . . . .

5. Look and read:

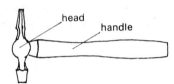

The head of the hammer *is fixed to* the handle. (It cannot move.)

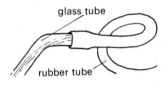

The rubber tube *is fitted over* the glass tube.

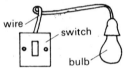

The wire *leads from* the switch *to* the bulb.

Now describe the following objects, using the words given:

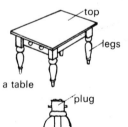

a table

consists of/fixed/supported/top . . . shaped

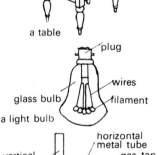

a light bulb

consists of/contains/shaped/attached/connected

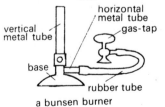

a bunsen burner

consists of/situated/connected/leads from . . . to/fitted over/fixed/shaped

21

6. Read this text:

> The apparatus for preparing hydrogen consists of a flask, a gas-jar, a beehive shelf, a trough, a delivery tube and a thistle funnel. The flask is spherical and has a flat bottom. It contains zinc and hydrochloric acid. The thistle tube and the delivery tube are fitted into the neck of the flask. They are held in place by a two-holed cork. The thistle tube leads down to the hydrochloric acid. The delivery tube leads from the flask to the hole in the beehive shelf. The beehive shelf is placed in the middle of the trough. The trough contains water. The gas-jar is supported by the beehive shelf. Hydrogen is collected at the top of the gas-jar.

Now draw the apparatus, using the parts shown in the diagram. The flask and the trough are in the right position. Then add labels to your diagram.

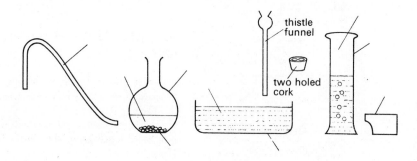

7. Now look at this diagram:

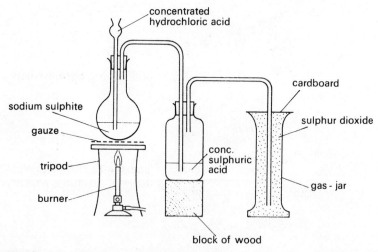

Apparatus for preparing sulphur dioxide

Describe the apparatus as fully as possible, using these words:

    consists of/situated/on the left/in the middle/on the right/supported by/placed under/there is . . . between . . . /fitted into/held in place by/contains/leads down into/leads from . . . to . . .

Section 3 Composition

8. Look at these examples:

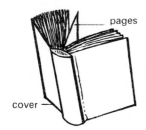

A book consists of pages and a cover.
It contains words and pictures.
The pages *are made of* paper and the cover is made of paper or cardboard.

Now make questions and answers about the following objects:

    Example: What does a hammer consist of?
                   It consists of a head and a handle.
                   What is it made of?
                   It is made of metal and wood.

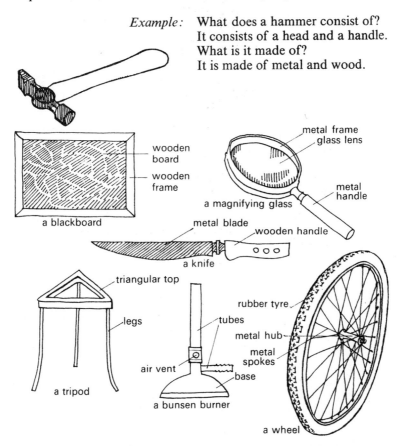

9. Look and read:

This square *is divided into* triangles.

The circle *is surrounded by* stars.

The milk bottle *is filled with* milk.

The body of a car *is covered with* paint.

Now answer these questions:

What is a chess-board divided into?
What is it made of?

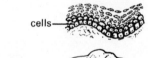

What is living matter divided into?

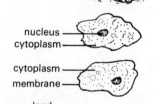

What is the nucleus of a cell surrounded by?

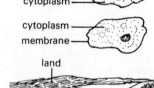

What is the cytoplasm covered with?

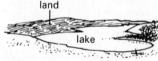

What is a lake surrounded by?
What is it filled with?

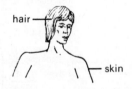

What are our bodies covered with?

10. Look at this diagram:

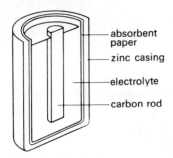

Now say whether these statements are true or false. Correct the false statements.

    a) The diagram shows a dry cell in cross-section.
    b) A dry cell is cylindrical in shape.
    c) The electrolyte is situated in the centre of the cell.
    d) The carbon rod is surrounded by the electrolyte.
    e) The thin layer between the casing and the electrolyte is made of paper.
    f) The cell is covered with absorbent paper.
    g) The biggest part of the cell is filled with electrolyte.

Unit A Revision

1. Look at this diagram:

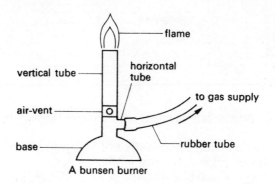

A bunsen burner

Now answer these questions:

Which parts are
        a) hollow?
        b) flexible?
        c) rigid?
        d) round?
        e) non-combustible?
        f) convex?
        g) circular?
        h) cylindrical?
        i) Which part leads to the gas supply?
        j) How is the vertical tube connected to the rubber tube?
        k) What shape is the flame?
        l) What are the tubes made of?
        m) Where is the air-vent?
        n) How is the rubber tube attached to the horizontal tube?
        o) What properties have the metal and rubber tubes in common?

2. Look at this diagram:

Now describe the diagram in sentences, using the following expressions:

a) at the top of
b) at the bottom
c) in the middle
d) on each side of
e) above
f) inside
g) below
h) over
i) diagonally above
j) in line with
k) inverted

3. Read this description and draw the diagram which it describes:
   At the top of the diagram there are two horizontal parallel straight lines. At the bottom there is a horizontal spiral. In the middle there is a circle. On each side of the diagram there is a cross. There are two inverted triangles diagonally above the circle, one on the left, the other on the right. The triangles are below the parallel lines. In each triangle there is a dot. Above the spiral and below the circle there is a square.

4. Read these short descriptions and then draw the objects and say what they are:
   a) The front of this object is circular or square in shape. Two arrow-shaped pointers are attached to the centre of the circle or square. The edge of the circle or square is divided into twelve equal sections. The front is usually covered with glass or plastic.
   b) This object consists of two flat parts, which are both made of wood, joined to each other at right angles. It is T-shaped.
   c) This object consists of a closed tube which contains a red or silver liquid, and is made of glass. The tube is cylindrical in shape, but one end is spherical in shape.
   d) This object consists of a beam (a horizontal bar), a vertical support under the middle of the beam and two hemispherical pans, which are attached to the ends of the beams by diagonal bars. At the top of the support, under the beam, there is a triangular fulcrum. The support is fixed to a flat base. In front of the support there is a pointer. The top of the pointer is attached to the beam. The tip is in front of a scale. When the two pans are in line the pointer is vertical and the tip of the pointer is in the middle of the scale.
   e) This object is made of wood, metal or plastic and is shaped like a long rectangle. The top and bottom edges are divided into sections by short vertical lines.
   f) This object is small and made of metal. One end is flat and circular in shape. There is a V-shaped groove across the middle of the circle. The other part of the object is at right angles to the flat end. It is tapering, with a pointed tip. Around the exterior of the tapering part there is a spiral-shaped groove.

Unit 4 Measurement 1

Section 1 Spatial measurements

1. Read this and replace the words in italics with expressions which you have already learnt:

This block of wood has various properties: for example, it is *shaped like a cube;* its material is wood; the material *burns easily; you cannot see through* it; the block is *difficult to bend,* etc.

2. Now read this:

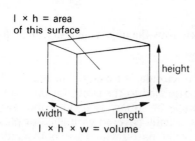

This block has other properties which are measured. It has *height, length* and *width.* Each surface has *area.* The area of the cross-section is the *cross-sectional area.* The area of all the surfaces is the *surface* area. The *volume* of the block = length x height x width (*equals* length *times* height *times* width).

Say which properties of these objects we can measure:

> *Example:* We can measure the radius, the diameter, the circumference and the area of a circle.

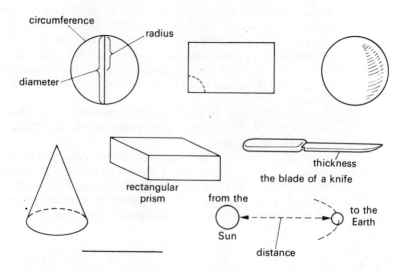

3. **Make sentences from the table below. Look in part 1 of the appendix for the names of the units of measurement which are given as abbreviations in the table.**

Example: The height of large objects is measured in metres.

The	height volume area width surface area length radius cross-sectional area diameter circumference	of	large small very small minute cylindrical	objects	is measured in	m cm mm μm m³ cm³ mm³ m² cm² mm²
	distance	between		places		km

4. **Complete these sentences:**

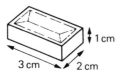

This brick has a length of 3 cm.
It has a _____ of 1 cm.
It has a _____ of 2 cm.
It has a _____ of 2 cm².
It has a _____ of 22 cm².
It has a _____ of 6 cm³.

Describe the measurements of this forest and the trees:

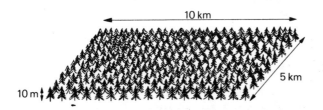

Complete this:

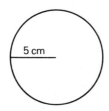

This circle has a radius of _____.
It has a diameter of _____.
It has a circumference of
_____. ($2\pi r$)
It has an area of _____. (πr^2)

Describe the measurements of this coin, including its thickness:

5. Look and read:

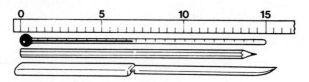

The thermometer has a length of *exactly* 15 cm.
The pencil has a length of *approximately* 15 cm. (exactly 14·9 cm)
The knife also has a length of approximately 15 cm. (exactly 15·2 cm)

Now complete these:

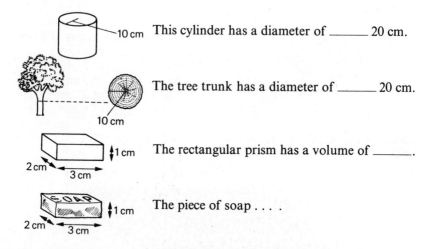

This cylinder has a diameter of _____ 20 cm.

The tree trunk has a diameter of _____ 20 cm.

The rectangular prism has a volume of _____.

The piece of soap . . . .

Section 2 Other measurements

6. Refer to part 2 of the appendix and say whether the following statements are true or false. Correct the false statements.

    a) Duration is measured in degrees Centigrade.
    b) The second is a unit of time.
    c) Speed is measured in kilograms per hour.
    d) The watt is a unit of electrical resistance.
    e) Density is measured in grams per metre cubed.
    f) The gram is a unit of mass.
    g) Liquid measurements are made in litres, or cubic decimetres.

7. Look at part 3 of the appendix and make questions and answers like the following:

> *Example:* What is the speed of light?
> The speed of light is 299,790 km/s.
>
> *or:* The speed of light is approximately 300,000 km/s.

Section 3 Scales and averages

8. Look at this diagram:

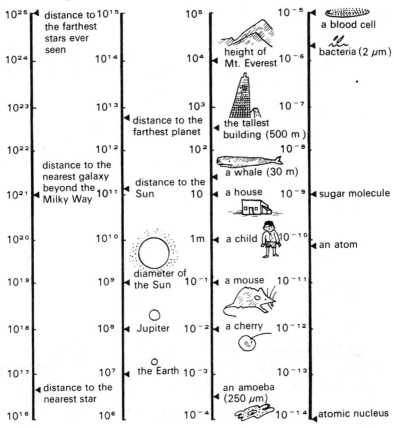

Scale of dimensions (measurements in metres)

Very large and very small quantities are expressed like this:
10^6 = ten *to the power of* six = one million.
10^{-6} = ten *to the power of minus* six = one millionth.

Complete these:
10^2 =   =
10^3 =   =
10^8 =   =
10^{-2} =   =
10^{-5} =   =

9. Refer to part 4 of the appendix and then make sentences like the example about these dimensions:

> the distance to the farthest stars, the diameter of the Sun, the diameter of the Earth, the height of Mount Everest, the length of a mouse, the diameter of a cherry, the diameter of a blood cell, the diameter of a sugar molecule

> *Example:* A mouse has a length of approximately ten to the power of minus one metres, ie ten centimetres.

10. Read this:

> The above diagram is a *scale* of dimensions. The dimensions *range from* the nucleus of an atom, which has a diameter of approximately 10^{-14} m, *to* the radius of the known universe, which is approximately 10^{25} m.

Now make sentences from the table:

| Plants Animals Buildings | range in size from | the whale, the tallest building, the tallest tree, | which | has a | height length | of approximately |

| 500 m, 80 m, 30 m, | to | the amoeba, houses, bacteria, | which | has have | a diameter a height | of approximately | 10 m. 250 μm. 2 μm. |

11. Look at these histograms:

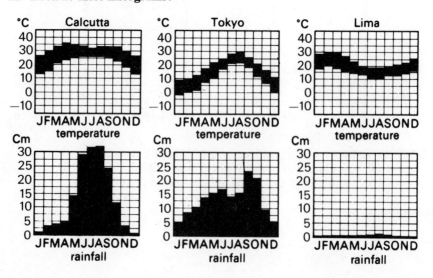

32

Now read this:

> The histograms in the top row show the average *range* of temperature (in degrees Centigrade) for each month in three cities. The histograms in the bottom row show their average monthly rainfall (in centimetres).
>
> In Calcutta in January the temperature ranges from 27°C to 13°C; that is, the *maximum* temperature is 27°C and the *minimum* temperature is 13°C. These are the two *extremes* of temperature.

Complete these (see part 5 of the appendix for the names of months):

> a) Extremes of temperature in Tokyo in January: maximum _____ ; minimum _____ .
> b) In Lima in April the temperature ranges from _____ to _____ .
> c) Throughout the year in _____ the rainfall ranges from 33 cm to 1 cm.
> d) In Tokyo the maximum rainfall occurs in the month of _____ , and the minimum rainfall . . . .

Now read this:

> The *average* monthly rainfall in Calcutta during the first six months of the year is:
>
> | | |
> |---|---|
> | January | 1 cm |
> | February | 3 cm |
> | March | 4 cm |
> | April | 5 cm |
> | May | 14 cm |
> | June | 28 cm |
> | Total | 55 cm ÷ 6 = 9·2 cm |

Answer these questions:

> e) Is the figure 9·2 exact or approximate?
> f) What is the total rainfall for the second half of the year in Calcutta?
> g) What is the average monthly rainfall during this period?
> h) What is the average monthly rainfall during the last three months of the year in Tokyo?

Now read this and answer the questions:

> In Lima the range of rainfall is very *narrow*. Rainfall is fairly *constant* throughout the year. In Calcutta, however, the range of rainfall is very *wide*. It *varies* a lot.
>
> i) In which city is there the widest range of temperature?
> j) In which city is the temperature most constant?
> k) Where does the rainfall vary most?

Appendix to Unit 4

Part 1 Units of measurement and their abbreviations

kilometre	km
metre	m
decimetre	dm
centimetre	cm
millimetre	mm
square metre	m^2
cubic metre (metre cubed)	m^3
micrometre	μm = ('micron')

formulae: The area of a circle πr^2
The circumference of a circle $2\pi r$

Part 2 Other measurements and their units

electric current	ampere (amp)
electric power	watt (W)
electric resistance	ohm (Ω)
electric potential difference	volt (V)
temperature	degrees Centigrade (°C)
mass	gram (g), kilogram (kg)
weight (the force of gravity on mass)	newton, kilonewton
speed	kilometres per hour (kph) (kmh^{-1})
density	kilograms per cubic metre (kg/m^3) (kgm^{-3})
time (duration)	second (s), minute (min), hour (hr)
fluid capacity	litre (l) = cubic decimetre (dm^3)

Part 3 Some facts

The speed of light is 299,790 kilometres per second (km/s) (kms^{-1})
The speed of sound in air is 332 (m/s) (ms^{-1})

Approximate height of mountains:	Everest	8,848 m
	Aconcagua	6,960 m
	Kilimanjaro	5,895 m
	Ararat	5,156 m
Approximate length of rivers:	Nile	6,656 km
	Amazon	6,480 km
	Missouri-Mississippi	5,936 km
	Yangtse	5,440 km

Areas and maximum depths of oceans (approximate):

	area	maximum depth
Pacific	102,177,600 km²	12,066 m
Atlantic	50,608,000 km²	9,166 m
Indian	29,996,000 km²	8,800 m

Boiling points:

water	100°C
oxygen	−183°C
alcohol	78°C

Freezing points:

water	0°C
oxygen	−218·4°C
alcohol	−131°C

Density:

gold	19,400 kg/m³
water	1,000 kg/m³
mercury	13,600 kg/m³
alcohol	800 kg/m³

Part 4 Prefixes used in units of measurement

kilo- = x one thousand: 1 km = 1000 m
deci- = one tenth: 1 dm = $\frac{1}{10}$ m $\qquad 10^{-1}$
centi- = one hundredth: 1 cm = $\frac{1}{100}$ m $\qquad 10^{-2}$
milli- = one thousandth: 1 mm = $\frac{1}{1000}$ m $\qquad 10^{-3}$
micro- = one millionth: 1 μm = $\frac{1}{1000000}$ m $\qquad 10^{-6}$
nano- = one thousand millionth: 1 nm = $\frac{1}{1000000000}$ m $\quad 10^{-9}$

Part 5 Months of the year

January	July
February	August
March	September
April	October
May	November
June	December

Unit 5 Process 1 Function and Ability

Section 1 Function

1. Look and read:

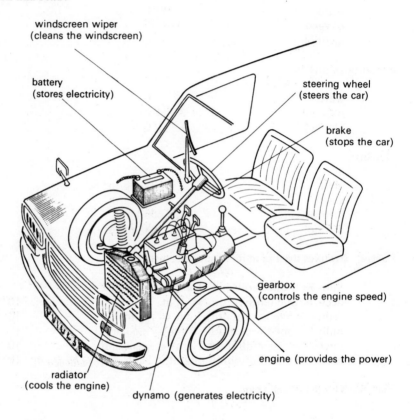

Parts of a car and their functions

Ask and answer questions:

    Examples: What *does* the gearbox *do*?
                  The gearbox *controls* the engine speed.

                  What *is* the gearbox *used for*?
                  The gearbox *is used for controlling* the engine speed.

2. Now write five sentences like the following:

    Example: The dynamo *serves to generate* electricity.

3. Look and read:

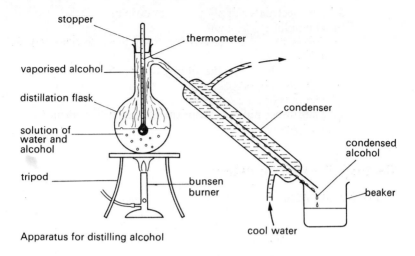

Apparatus for distilling alcohol

Now make true sentences from this table:

The function of	the thermometer the stopper the bunsen burner the tube at the top of the condenser the cool water the tube at the bottom of the condenser the beaker the tripod	is to	conduct cool water in. heat the liquid. collect the condensed alcohol. measure the temperature of the solution. conduct the cool water away. support the apparatus. cool the vaporised alcohol. hold the thermometer.

4. Look and read:

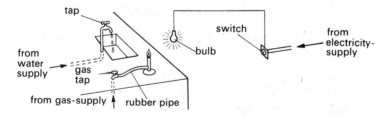

The water-supply *provides* water.
The water-pipe *conducts* water from the water-supply to the tap.
The tap *controls* the supply of water.

Now complete these sentences:

a) The electricity supply _____ electricity.
b) The bulb _____ light.
c) The wire _____ electricity from the electricity-supply to the light.
d) The switch _____ the supply of electricity.
e) The gas-supply _____ gas.
f) The burner _____ heat.
g) The rubber pipe ... from the gas-tap to the burner.
h) The gas-tap ....

Section 2 Instruments

5. Look at this picture:

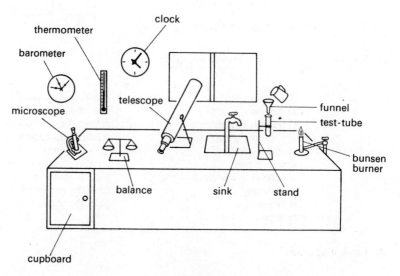

Now complete these sentences by adding the names of objects in the picture:

Example: A clock is an instrument for measuring time.

a) ... an instrument for measuring temperature.
b) ... a device for looking at distant objects.
c) ... a place for washing things.
d) ... a device for supplying heat.
e) ... an instrument for weighing things.
f) ... a device for holding chemicals during experiments.
g) ... a place for storing things.
h) ... an instrument for looking at very small things.
i) ... a device for pouring liquids.
j) ... an instrument for measuring atmospheric pressure.
k) ... a device for controlling the supply of water.
l) ... a device for supporting things during experiments.

6. Look and read:

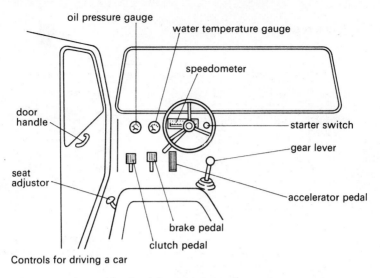

Controls for driving a car

The controls *enable* the driver *to drive* the car.

This means:

With the help of the controls, the driver *can* drive the car.

Now make similar sentences from this table:

The door handle		check the temperature.
The steering wheel		stop the car.
The seat adjustor		steer the car.
The gear lever		check the oil pressure.
The accelerator pedal	enables	change gear.
The clutch pedal	the	start the engine.
The brake pedal	driver	check the speed.
The oil pressure gauge	to	get in and out.
The water temperature gauge		control the speed.
The speedometer		adjust the seat.
The starter switch		operate the gear lever.

Section 3 Ability and capacity

7. Look and read:

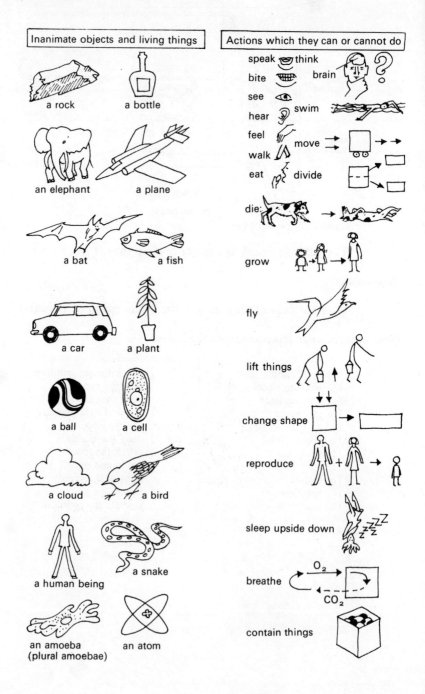

Now make questions and answers like the following:

> *Examples:* Can a snake breathe?
> Yes, a snake can breathe.
>
> Can a snake speak?
> No, a snake cannot speak.

8. Make true sentences from this table:

All	animals living things inanimate objects	can	swim. fly. divide.
Some	human beings cells		reproduce. change shape.
No	plants		speak.

9. Look at these examples:

> Plants *can grow.* = Plants *are able to grow.*
> = Plants *have the ability to grow.*
> = Plants *have the capacity to grow.*
> = Plants *are capable of growing.*

Now write full sentences in answer to these questions:

a) What living things are capable of flying?
b) What animals have the capacity to sleep upside down?
c) What living things can swim?
d) What inanimate objects have the capacity to fly?
e) What living things are capable of feeling?
f) What living things have the ability to think?
g) What inanimate objects are capable of changing shape?
h) What living things are able to reproduce?
i) What living things cannot walk?
j) What inanimate objects are capable of moving?

Section 4 Functions in the human body

10. Look and read:

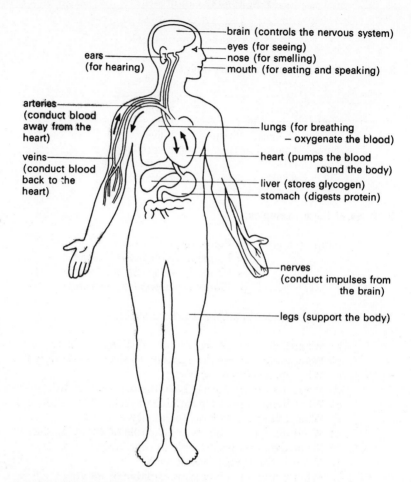

Internal and external parts of the human body and their functions

Now complete these sentences:

a) Our eyes _____ us _____ see.
b) With the _____ of our mouths we are _____ to speak and eat.
c) Our ears are organs for _____.
d) With the help of our noses we _____ smell things.
e) Our lungs enable us to _____.
f) The lungs serve to . . . .
g) The _____ of the heart is to circulate the _____.
h) The heart acts as a _____ for the blood.
i) The stomach is used for . . . .

j) The liver is a place for _____ _____.
k) The _____ act as a support for the body.
l) The function of the nerves is to . . . .
m) The brain is used for . . . .
n) The veins . . . .
o) The arteries serve . . . .

Unit 6 Process 2 Actions in Sequence

Section 1 Preceding, simultaneous and following events

1. Look and read:

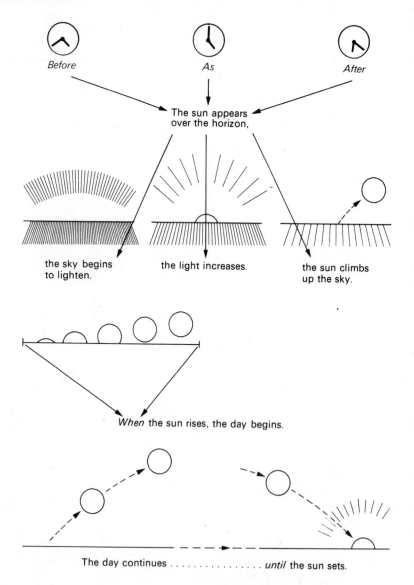

Each sentence contains two events, X and Y. Read out the sentence which means:

a) X occurs at the same time as Y (simultaneously with Y).
b) X occurs at approximately the same time as, or soon after, Y.
c) X precedes Y.
d) X follows Y.
e) Y is at the end of X.

Now complete these sentences:

f) _____ the sun rises, the air temperature rises.
g) _____ the sun reaches the highest point in the sky, it begins to descend.
h) _____ the sun descends, the air temperature falls.
i) _____ the sun sets, it approaches the horizon.
j) _____ the sun sets, the sky becomes completely dark.
k) _____ the sun sets, the day ends.
l) The night begins _____ the sun sets.
m) The night continues _____ the sun rises.

2. Number these events in the order in which they occur when water is heated. Give simultaneous actions the same number.

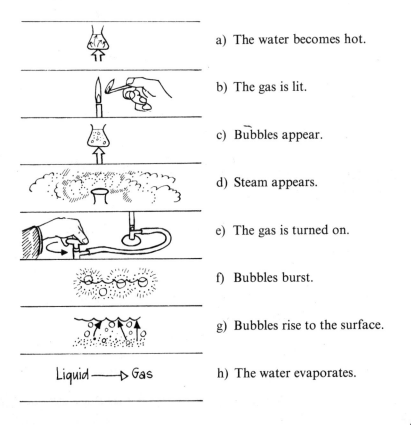

a) The water becomes hot.

b) The gas is lit.

c) Bubbles appear.

d) Steam appears.

e) The gas is turned on.

f) Bubbles burst.

g) Bubbles rise to the surface.

h) The water evaporates.

Now look at the two points in this example:

> *As soon as* the gas is turned on, *it* is lit.
> (X follows Y immediately)

Complete these sentences:

    i) As the water evaporates, ...
    j) As soon as the bubbles burst, ...
    k) When the bubbles rise to the surface, ...
    l) As soon as the gas is turned on, ...
    m) Before the gas is lit, ...
    n) After the water becomes hot, ...
    o) As soon as the bubbles appear, ...

3. Number these events in the order in which they occur:

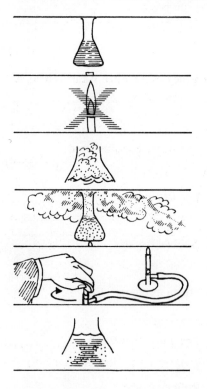

    a) The water ceases boiling.

    b) The flame is extinguished.

    c) The water starts to boil.

    d) The water continues boiling.

    e) The gas is turned off.

    f) The bubbles disappear.

Now write complete sentences joining these pairs of events and making any other necessary changes:

d + *b	*e + b	*Use one of these expressions:		
*c + d	*a + e	when	before	as soon as
*b + a	*a + f	until	after	

Example: As soon as the gas is turned off, the flame is extinguished.

Section 2 Sequences

4. Look at this diagram:

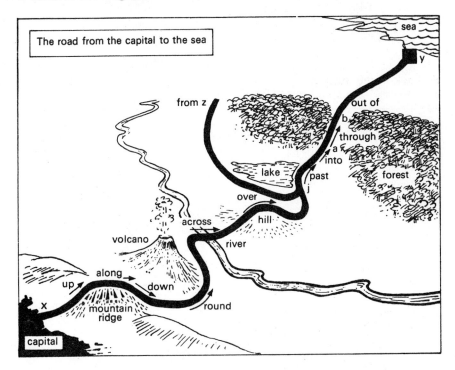

Now complete this description:

A road leaves the capital at **x**. It goes _____ a mountain, _____ a ridge and _____ the other side. It goes _____ a volcano, _____ a river and _____ a hill. It joins the road _____ **z** _____ the junction, **j**. It goes _____ a lake and _____ a forest. It goes _____ the forest _____ **a** and comes _____ the forest _____ **b**. The road reaches the sea _____ **y**.

Look at these sentences:

After *the road leaves* the capital at **x**, *it* goes up the mountain.
After leaving the capital at **x**, *the road* goes up the mountain.

When the road reaches **a**, it goes into the forest.
On reaching **a**, the road goes into the forest.

Change these sentences into the second form:

a) After the road goes round the volcano, it crosses the river.
b) Before the road enters the forest, it passes the lake.
c) When the road emerges from the forest, it is near the sea.
d) After the road ascends the hill, it goes along the ridge.

47

e) The road continues until it reaches the sea.
f) Before it goes round the volcano, the road descends the mountain.
g) As it goes between the hill and the lake, the road joins the road from **z**. (*While* going . . .)
h) When the road reaches the volcano, it goes round it.
i) As the road travels from **x** to **y**, it crosses a river.

5. **Look at this diagram:**

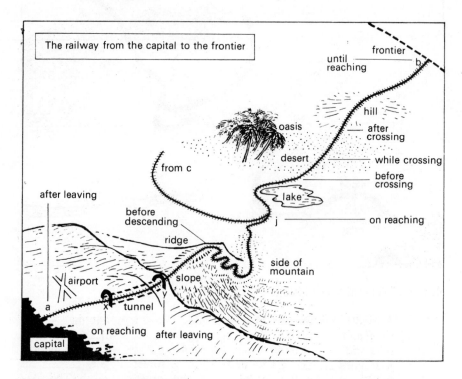

Now write nine sentences using the opening phrases in the diagram.

> *Example:* After leaving the capital at **a**, the railway goes past the airport.

6. **Look at these sentences:**

> *While crossing* the desert, the railway *passes* an oasis.
> The railway *crosses* the desert, *passing* an oasis.
> (X is simultaneous with Y, or occurs during Y)

Now change these sentences into the second form:

a) While going through the tunnel, the railway passes under the mountain.
b) While descending the mountain, the railway makes several turns.

c) While passing the lake, the railway travels in a semi-circle.
d) While crossing the desert, the railway passes an oasis.
e) While approaching the frontier, the railway goes over a hill.
f) While travelling between the mountain and the lake, the railway joins the railway line from **c**.

Section 3 Cycles

7. Look and read:

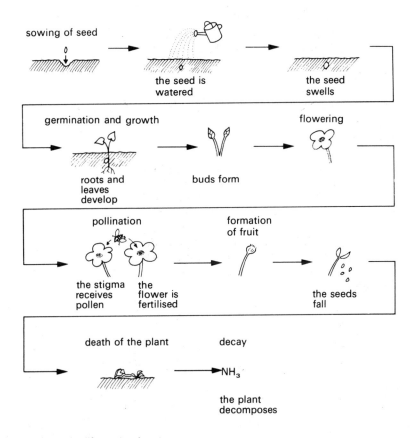

Stages in the life-cycle of a plant

Look at these examples:

Preceding actions:
Before the plant *germinates*, it is watered.
Before ⎫
Prior to ⎭ *germination*, the seed is watered.

Following actions:
After the plant *germinates*, the roots and leaves develop.
After germination, the roots and leaves develop.

Simultaneous actions:
As the plant *germinates*, the seed swells.
During germination, the seed swells.

And this example:

After the seed is watered, *germination* ⎧ *occurs.*
⎩ *takes place.*

Now answer these questions:

a) What happens prior to germination?
b) What occurs during growth?
c) What happens before flowering?
d) What takes place after pollination?
e) What happens after the seeds fall?
f) What occurs before the plant decomposes?
g) What occurs as the plant decomposes?

8. Look at this:

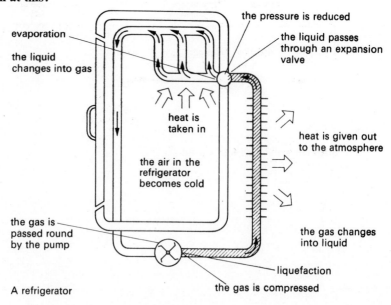

A refrigerator

Write a description of the cycle of events in a refrigerator.

9. Look and read:

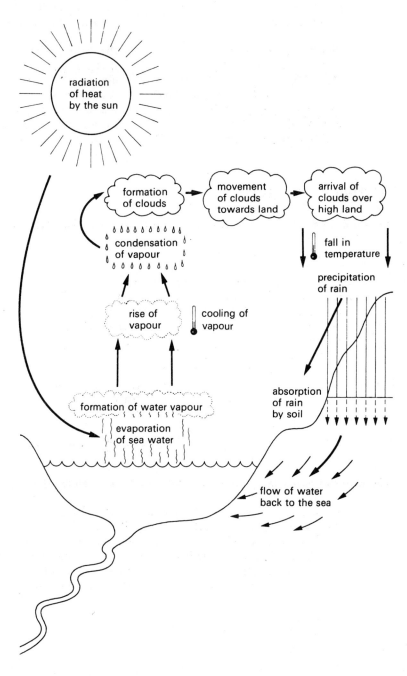

The water cycle

Now write a description of the cycle by joining the correct half-sentences:

When the sun radiates heat,   until they reach high land.
As soon as the water vapour forms,   rain is precipitated.
While rising,   the water flows back to the sea.

When the vapour cools,   the vapour cools.
During condensation,   it begins to rise.
The clouds then move towards land,   clouds are formed.
When the clouds reach high land,   it condenses.
As the temperature falls,   sea water evaporates.
On being precipitated,   the air temperature falls.
After being absorbed,   the rain is absorbed by the soil.

Section 4   Stages

10. Look again at the life-cycle of a plant and then read this description:

First, the seed is sown.
  Next, it is watered.
    Then, the seed begins to swell.
      At this stage, germination begins.
        Subsequently, the roots develop.
        Meanwhile, the leaves also develop.

Later, flowers appear.
  Then, pollination takes place.
    During this process, the stigma receives pollen.
      Afterwards, the fruit forms.

      Eventually, the plant dies.
        Finally, the plant decomposes.

The words in *italics* mark stages in a process. Now give the following:

  a) A word which marks the opening, or initial stage.
  b) A word which marks the last, or ultimate, stage.
  c) Two words which mark next or following stages
  d) Three expressions which mark events occurring some time later.
  e) Three expressions which mark simultaneous events.
  f) One word which marks an event occurring after a long process.

11. Put these stages in the right order and then match them with the expressions on the left:

    Example: First, the site is bought.

Stages in building a house

        First,   the drains are dug.
        Then,   the materials are bought.
        Meanwhile,   the house is painted.
        Subsequently,   the walls are built.
        At this stage,   the site is bought.
        Next,   the site is levelled.
        Afterwards,   the foundations are laid.
        Then,   the house is ready to live in.
        Later,   the roof is made.
        Eventually,   the doors and windows are put in.
        Finally,   the electricity and water systems are installed.

Unit B Revision

1. Look at this diagram and read the passage which follows:

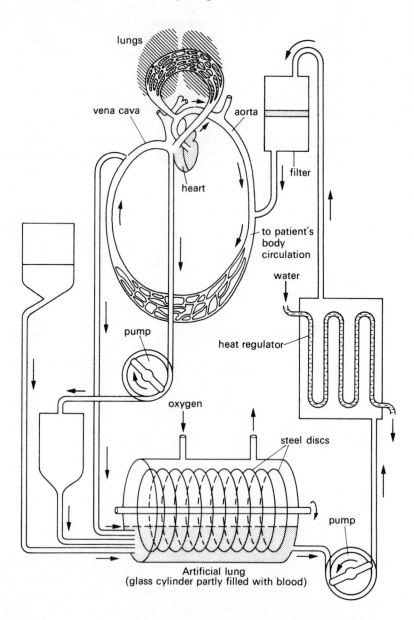

The heart-lung machine

The heart-lung machine is used for maintaining the circulation and oxygenation of the patient's blood. It consists of an artificial lung, pumps, tubes and devices for controlling the heat and filtering the blood. The artificial lung serves to oxygenate the blood, which is diverted from the vena cava before reaching the heart.

On leaving the vena cava, the blood enters a plastic tube and flows down this until it enters the artificial lung. This is a horizontal glass cylinder which is partly filled with blood. It contains rotating steel discs. After the blood enters the cylinder it forms a thin film on the surface of the discs. This enables the blood to absorb oxygen, which is pumped through the cylinder. The oxygenated blood subsequently passes through a heat regulator and a filter before returning to the patient's body circulation.

Now answer these questions:

a) What does the artificial lung do to the blood?
b) What is the function of the heat regulator?
c) What does the artificial lung consist of?
d) Why is the blood able to absorb oxygen?
e) Look again at paragraph 2, and complete this chart showing stages in the movement of blood through the heart-lung machine.

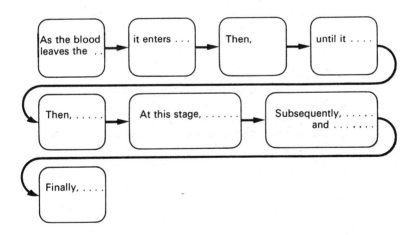

2. Read this passage and answer the questions:

Atmospheric pressure

The average atmospheric pressure at sea-level is approximately 1 kg/cm². This is the pressure which will support a column of mercury (Hg) 760 mm high, and it is called 'one atmosphere'. It is equivalent to the pressure of a column of air approximately 8 km high, if the density of this air is constant and equal to the pressure at sea-level.

a) What is pressure measured in?
b) What is the unit of atmospheric pressure?
c) What is the average pressure of the atmosphere at sea-level?
d) Is atmospheric pressure constant at sea-level or does it vary?
e) How is one atmosphere defined?

3. **Read the following paragraph and answer the questions below it:**

A barometer is a device for measuring atmospheric pressure. Modern barometers are usually of the aneroid type. This consists of a sealed metal box which has no air in it. The top and bottom of this box are thin plates, which are partly curved and partly flat. In the interior of the box there are springs, which push against the top and bottom plates against external air pressure. When this decreases, the springs can expand but when it increases the springs are compressed. The barometer also contains a pointer joined to the top plate which serves to indicate the change in pressure.

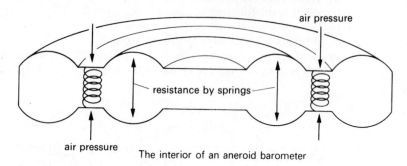

The interior of an aneroid barometer

a) What is the function of a barometer?
b) What does the sealed metal box contain?
c) What is the function of the springs?
d) What happens when the atmospheric pressure increases?
e) Which part of the barometer enables us to know the atmospheric pressure?

Unit 7 Measurement 2 Quantity

Section 1 How much and how many?

1. Look at these diagrams:

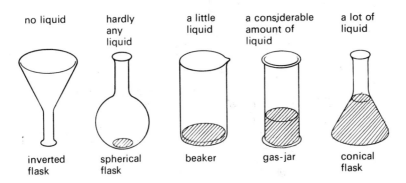

Make questions and answers like the following:

>   *Example:* How much liquid does the beaker contain?
>   It contains a little liquid.

Now look at this:

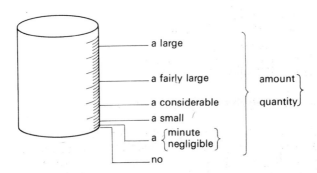

>   **Note:** *Considerable* means large enough to be important.
>   *Negligible* means too small to be important.
>
>   A lake contains a *very large* quantity of liquid.
>   The sea contains an *enormous* amount of liquid.

Now write sentences describing the quantities of liquid in the containers above.

2. Look at this:

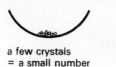
a few crystals
= a small number
of crystals

a considerable
number of crystals

many crystals
= a large number
of crystals

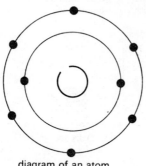

diagram of an atom
of oxygen

Oxygen has 8 electrons. This is a fairly small number.

Give the names of elements which have, in one atom:

a few electrons
a large number of electrons
a very large number of electrons
a fairly large number of electrons
a very small number of electrons

3. Complete these sentences:

a) How _____ electrons does an atom of sodium possess?
b) How _____ oxygen does the atmosphere contain?
c) Our bodies contain a very large _____ of water.
d) A large _____ of whales are found in the Pacific Ocean.
e) There is an _____ number of stars in the universe.
f) For a rich man, one dollar is a _____ quantity of money, but for a poor man it is a _____ amount.
g) The air consists of _____ nitrogen (78%), _____ oxygen (21%), _____ argon (less than 1%) and _____ helium, neon, krypton and xenon.
h) An orange contains _____ seeds.
i) The _____ of bacteria in the air is _____.
j) _____ people can speak more than 5 languages.

4. **Look again at the diagrams of the containers (exercises 1 and 2) and read this:**

   The conical flask contains *much more* liquid than the beaker.
   The beaker contains *considerably less* liquid than the gas-jar.
   The beaker contains *slightly more* liquid than the spherical flask.
   The dish on the right contains *many more* crystals than the dish on the left.
   The dish on the left contains *considerably fewer* crystals than the dish in the middle.

Complete these:

   a) The conical flask contains ... than the gas-jar.
   b) The spherical flask contains ... than the beaker.
   c) The beaker contains ... than the conical flask.
   d) The dish in the middle contains ... than the dish on the right.
   e) The dish on the right contains ... than the dish in the middle.

5. **Look at these tables:**

   Elements in the Earth's crust:

		%
Oxygen	(O)	49
Silicon	(Si)	26
Aluminium	(Al)	8
Iron	(Fe)	5
Calcium	(Ca)	3*
Sodium	(Na)	3*
Potassium	(K)	2
Magnesium	(Mg)	2
Titanium	(Ti)	0·63
Hydrogen	(H)	0·13
Phosphorus	(P)	0·13
Manganese	(Mn)	0·10
Sulphur	(S)	0·052
Carbon	(C)	0·032

   *approximately equal quantities

   Number of electrons in one atom of some metals:

Lithium	(Li)	3
Sodium	(Na)	11
Magnesium	(Mg)	12
Potassium	(K)	19
Manganese	(Mn)	25
Iron	(Fe)	26
Copper	(Cu)	29
Zinc	(Zn)	30
Strontium	(Sr)	38
Tin	(Sn)	50
Gold	(Au)	79
Mercury	(Hg)	80
Lead	(Pb)	82
Radium	(Ra)	88
Uranium	(U)	92

Now make comparisons like the following examples:

   The Earth's crust contains much more oxygen than magnesium.
   An atom of iron possesses slightly fewer electrons than an atom of copper.

Section 2 Enough and too much

6. Look at this:

These are the *average quantities* of food consumed by *1 person* at the Asia Restaurant:

Meat	125 grams
Rice	150 grams
Bread	100 grams
Onion	1
Tomatoes	2

If the restaurant has the following quantities, calculate how much of each kind of food is available:

Meat	6 kilos
Rice	9 kilos
Bread	5 kilos
Onions	60
Tomatoes	80

Example: There is *enough meat for* 48 people.

a) (Rice)
b) (Bread)
c) (Onions)
d) (Tomatoes)

7. Say whether the supply of each kind of food is adequate or not when exactly **50 people** eat at the restaurant.

Example: There is *exactly enough* bread.

a) There is *too much* _____.
b) There are *too many* _____. } (ie more than enough)
c) There is *too little* _____.
d) There are *too few* _____. } (ie not enough)

8. Now look at this:

One day, 55 people come to the restaurant, which has these quantities of food:

Meat	7 kilos
Rice	8·25 kilos
Bread	5 kilos
Onions	50
Tomatoes	120

Write down how *much* (or *many*) of each kind of food there *is* (or *are*) using these words:

    a) too little
    b) too few
    c) exactly enough
    d) too much
    e) too many

9. Look at these examples:

To be healthy, you must eat { the right quantity of / enough / a sufficient amount of / an adequate amount of } food,

but { too much / an excess of / an excessive amount of } food makes you fat,

and { too little / a lack of / an insufficient quantity of / an inadequate amount of } food makes you thin, and hungry.

From your own knowledge, can you say which is the right phrase in each of these sentences?

    a) (Too much/a lack of/an excess of) iron causes anaemia.
    b) (A lack of/a sufficient quantity of/an excess of) carbohydrate causes fatness.
    c) If your food has not (sufficient/excessive/insufficient) calories, you will not have enough energy.
    d) (The right quantity/an inadequate amount/an excess) of vitamins is necessary for health.
    e) (An adequate amount/an excessive amount/an inadequate amount) of calcium causes bone disease.
    f) If you have (an insufficient amount/an excessive amount/a sufficient amount) of clothing, you will be too hot.
    g) If you have (an adequate amount of/an excess of/too little) clothing, you will be too cold.
    h) Unless they have (a lack of/ an adequate amount of/an excessive amount of) water, plants will not grow.

10. Look at this example:

Why can't you take a photograph? (light)
Answer: Because the light is insufficient.

Answer the following questions, using these words:

    insufficient, inadequate, sufficient, adequate, excessive.

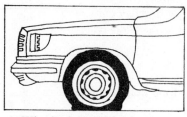

a) Why is the tyre flat?
(air pressure)

e) Why can't the boat go further?
(water)

b) Why has the tyre burst?
(air pressure)

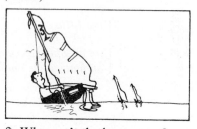

f) Why can't the boat move?
(wind)

c) Why does the light shine brightly?
(current)

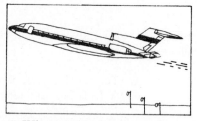

g) Why can the plane leave the ground?
(speed)

d) Why will the boat sink?
(cargo)

Section 3 Too small and big enough

11. Look at these examples:

> Why can't you write your name in this rectangle? ☐
> *Answer:* Because it's *too small to write in*.
>
> Why can you write it here? ☐
> *Answer:* Because it's *big enough to write in*.

Now answer these questions (using the words in *italics*):

a) Why can't you *put* an elephant *in your pocket?*

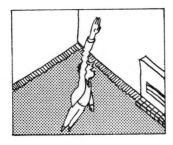

b) Why can't you *touch* the ceiling?

c) Why can't you *lift* a lorry?

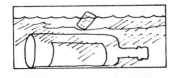

d) Why does a cork *float?*

e) Why can't the lorry *go down the street?*
(wide)

 f) Why can the car *go down the street*? (narrow)

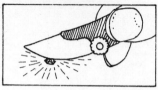 g) Why can't you *cut* a diamond?

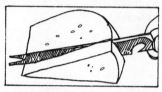 h) Why can you *cut* cheese?

 i) Why can you *bend* rubber? (flexible)

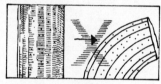 j) Why can't you *bend* concrete? (rigid)

 k) Why can you *see bacteria* with a microscope? (powerful)

 l) Why can't you *see bacteria with* a magnifying glass? (weak)

 m) Why can't you *see through* paper? (opaque)

12. Look at this map:

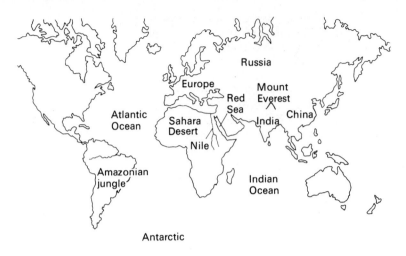

Look at these examples:

> Europe is *too cold for* tea plants.
> Parts of India are *warm enough to grow tea in.*

Now write ten sentences like these examples with *too/enough* and *for* (+ noun) or *to* (+ verb), using these words:

the Antarctic the Sahara Desert Mount Everest the Atlantic Ocean the Red Sea Russia the River Nile the Nile valley the Amazonian jungle Northern Europe China etc.	cold warm hot dry damp deep big wide narrow salty fertile infertile temperate high	for	penguins flowers camels tea plants bananas apples people fresh water fish wheat
		to	walk across live in (or on) visit in one day grow many plants in drink irrigate a large area of land support a large population swim across

13. Look at the map again and answer these questions:

    a) Where is there a lack of oxygen?
    b) Where is there an excess of salt?
    c) Where is the land *over*-populated?
       (ie the population is excessive)
    d) Where is the land *under*-populated?
       (ie there is a lack of population)
    e) Where is there a lack of vegetation?
    f) Where is the cold excessive?
    g) Where is the rainfall inadequate?

Unit 8   Process 3   Cause and Effect

Section 1   Actions and results

1. Look at this example:

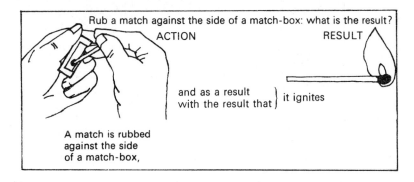

Now make similar statements about the actions and results below (using *so that*):

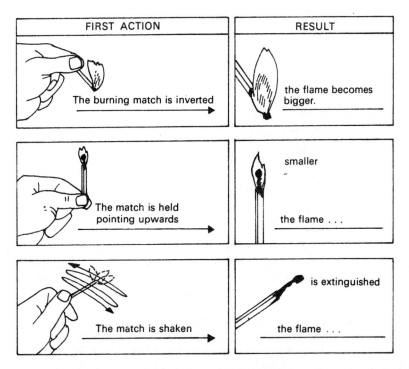

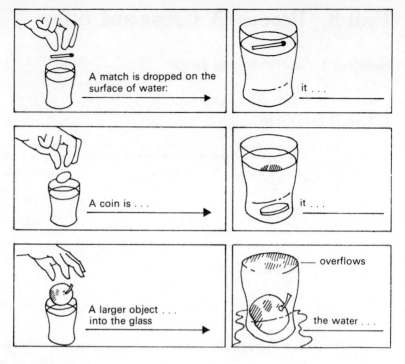

2. Look at this example:

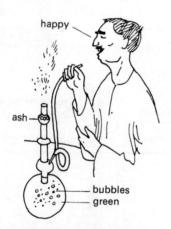

Changes of state

The process of smoking a water-pipe: smoke is sucked down the pipe, and as a result the smoke *changes into* bubbles, the water *turns* green and the smoker *becomes* happy (or ill). *Another result is that the tobacco is converted into* ash.

Make statements about the following actions and resulting changes, using *with the result that* or *and as a result*, and the words *become* (+ adjective), *turn* (+ colour), *change into* or *be converted into* (+ noun).

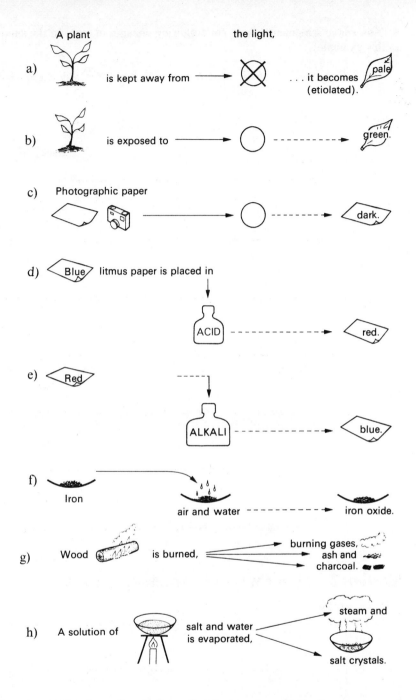

3. Now change the above descriptions of actions and results in the same way as the example:

> *Example:* *If* a plant *is kept* away from the light, it *will become* etiolated.

4. Now make statements about the following changes of state in the same way as this example:

Example: If ice is heated to melting point, it will melt, *changing* into water.

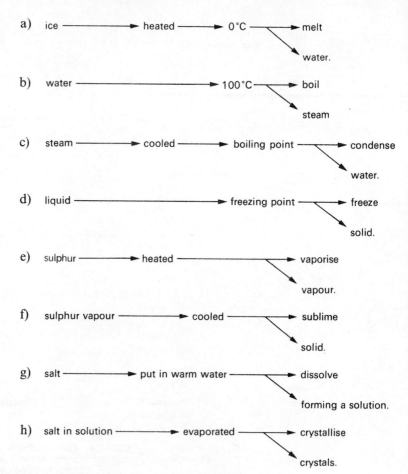

Section 2 Other ways of expressing results

5. Look at this:

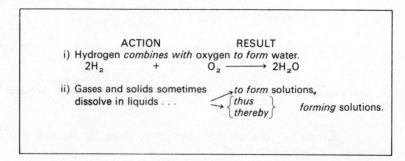

Make sentences describing chemical reactions from this table:

Potassium Calcium Magnesium Iron Carbon	combines with	hydrogen oxygen chlorine iodine	to form	calcium oxide, CaO. methane, CH$_4$. potassium iodide, KI. iron (III) oxide, Fe$_3$O$_4$. magnesium chloride, MgCl$_2$.

6. Read these statements about the action of chemical substances and then complete the examples which follow them:

    Example: Elements *combine to make* compounds.
                For example, sodium combines with chlorine to form sodium chloride.

    a) Solids may burn in gases *to form* different solids.
        For instance, calcium ... oxygen ... calcium oxide, CaO.

    b) Some substances absorb others, *thus producing* other substances.
        _____, chlorophyll _____ water and carbon dioxide, ... oxygen and sugars.

    c) Gases and solids sometimes dissolve in liquids, *thereby producing* solutions.
        Thus, sulphur dioxide _____ in water, ... sulphurous acid, H$_2$SO$_3$.
        Another example is hydrogen chloride, which ... , ... hydrochloric acid, HCl.

    d) Radioactive substances disintegrate *to form* more stable elements.
        _____, thorium ... lead 208.
        ... is proactinium, which ... lead 207.

    e) Metals often react with acids *to give* metal salts.
        For example, zinc ... sulphuric acid ... zinc sulphate, ZnSO$_4$.

    f) An acid is neutralised by an alkali *to form* a salt.
        _____, hydrochloric acid ... sodium hydroxide solution ... sodium chloride.

    g) Calcium reacts with water, *thereby liberating* hydrogen and *producing* calcium hydroxide.
        Similarly, sodium ... , ... hydroxide.

    h) Calcium carbonate is decomposed by heating *to produce* calcium oxide and carbon dioxide.
        _____, potassium chlorate ... potassium chloride and oxygen.

7. Read this passage and look at the diagram:

The carbon cycle
The life of plants and animals depends on chemical substances containing carbon atoms. Plants obtain carbon from the very small amounts of carbon dioxide in the atmosphere. This atmospheric CO_2 is continually absorbed and given off (released) in the 'carbon cycle'.

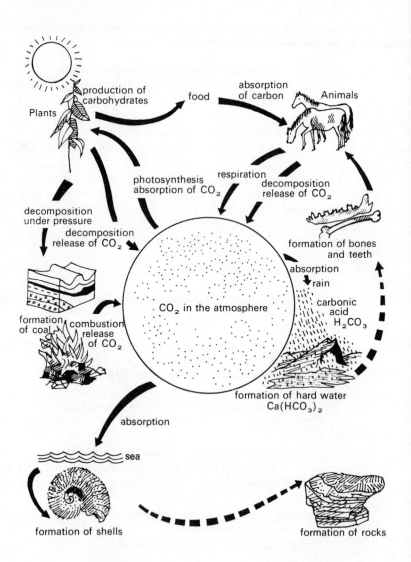

Look at these:

(A = cause, B = result)

A *results in* B.
B *results from* A.
As a result of A, B occurs.
A *leads to* B (eventually: other events occur between A and B).

Now make ten true sentences from the tables below:

As a result of	eating plants, photosynthesis, combustion of coal, decomposition of dead plants,	carbon dioxide is given off. carbohydrates are produced by plants. animals absorb carbon.

Decomposition of plants under pressure Release of CO_2 into the atmosphere Decomposition of dead animals Formation of hard water Absorption of CO_2 by the sea Production of carbohydrates Formation of carbonic acid Formation of shells	results in results from leads to	respiration. photosynthesis. the formation of teeth and bones in animals. the formation of rocks. the formation of coal. the release of CO_2 into the atmosphere. the formation of shells. the combination of rain and CO_2 in the atmosphere.

Section 3 Causing, allowing, preventing

8. **Read this sentence and look at the diagrams:**

   A valve *allows* liquid *to flow* one way and *prevents* it *from flowing* the other way.

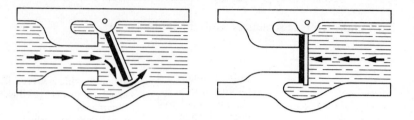

Answer these questions:

   a) Can the liquid flow from left to right?
   b) Why?
   c) Can the liquid flow from right to left?
   d) Why not?

Look at the diagram of the heart and make two sentences about the function of the valves, using the words given:

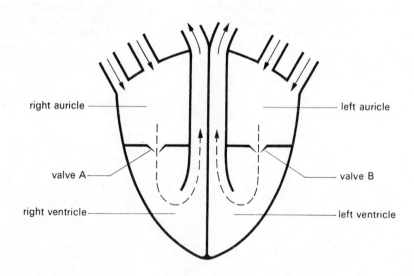

   e) Valve A allows blood/from/to/but prevents/from/to.
   f) Valve B . . . .

9. Read these sentences:

Sucking air up a tube *causes* the air pressure *to decrease*, with the result that it becomes lower than the atmospheric pressure.

Atmospheric pressure *consequently makes* the water *enter* the tube

Now describe the action of a syringe:

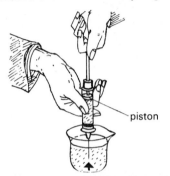

Pulling back the piston. . . .

10. Look at this diagram of a lift pump:

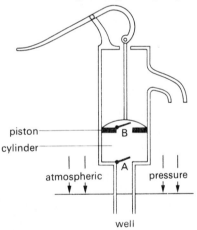

Now make sentences from this table:

The result of raising the piston	from returning to the well.
Atmospheric pressure makes the water	is that the water is compressed.
Valve A allows the water	to pass the piston.
Valve A prevents the water	enter the cylinder.
The result of lowering the piston	to enter the cylinder.
Compression makes the water	from returning past the piston.
Valve B allows the water	is that the air pressure is reduced.
Valve B prevents the water	pass the piston.

11. **Look and read:**

   The effects of heat
   These are observable changes produced by heating; some of them help chemists to identify substances:

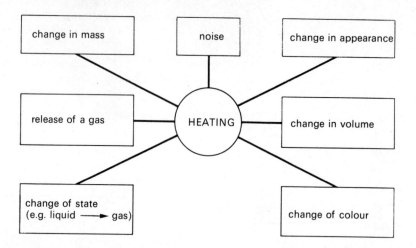

Now make pairs of sentences like the following. Use your own knowledge of chemistry, or use previous examples from this unit, or think of the effects of cooking:

$$\textit{Example:} \text{ Heating} \begin{cases} \textit{produces} \\ \textit{brings about} \\ \textit{causes} \end{cases} \text{a change of state in ice.}$$

It *causes it to* melt.

Section 4 Explanations

12. **Look at this example:**

   Why do matches ignite when rubbed?

   Matches ignite when rubbed $\rightarrow$ $\begin{cases} \textit{because of} \\ \textit{owing to} \end{cases}$ $\rightarrow$ friction.
   $\searrow$ *because* friction generates heat.

Now write explanations in answer to the following questions. The sentences and phrases on the right will help you. Choose between *because* and *because of/owing to*.

Note: Do not repeat the noun after *because*; use a pronoun (it/they)

Example: Wood floats on water because *it* is less dense than water.

Questions	Reasons/causes
Why do matches ignite when rubbed?	Wood is less dense than water.
Why are iron objects attracted or repelled?	Metal is denser than water.
Why do plants look green?	Plants contain chlorophyll.
Why does a liquid in a tube look like this?	Friction produces heat.
	Plants use carbon dioxide and water to produce carbohydrates.
Why does wood float on water?	Air is denser than hydrogen.
Why does water enter a pump?	a difference in air pressure
Why does blood flow only one way in the heart?	valves
	magnetism
Why do hydrogen balloons rise in air?	surface tension
Why are some rocks used in making cement?	gravity
	the calcium carbonate present in some rocks
Why do objects fall to the ground?	
Why do metal objects sink in water?	
Why do plants need CO_2?	

13. Now make some statements like the following pair of sentences, using *therefore* **or** *consequently*:

Example: Plants contain chlorophyll. *Therefore, Consequently,* } they look green.

Unit 9 Measurement 3 Proportion

Section 1 Relative size

1. Look and read:

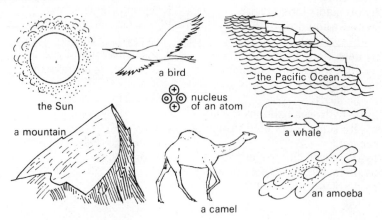

The things in the picture are not drawn *to scale*. The Sun is in fact very much bigger than it appears in the picture. The nucleus of an atom is very much smaller.

Sizes are *relative*. Most objects are big *in proportion to* the size of an atom but small in proportion to the size of the Sun.

Make sentences like the following:

> *Example:* A camel is big in proportion to the size of an amoeba but small in proportion to the size of a mountain.

2. Now make questions and answers like the following:

> *Example:* Is a mountain large or small?
> *Compared with* the size of the Sun, a mountain is *relatively* small.

3. Read this:

> The length of the river Nile: 6,656 km
> The height of Mount Everest: 8,848 m
> The population of New York: 11,528,649

Now make sentences about rivers, mountains and cities in your country, like this:

Compared with (river), (river) is relatively { high,
(mountain), (mountain) long,
(city), (city) has a relatively large population,

but compared with { the Nile, Everest, New York, } it is relatively { low. short. small. }

4. Now look at this bar-graph:

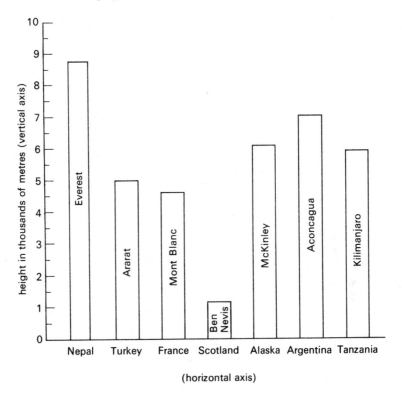

The relative heights of Mountains

Read and complete this:

The rectangles show the heights of different mountains.
The heights are marked on the _____ axis. The scale is shown in thousands of _____. The heights of these mountains range from _____ to _____.

Now compare the heights of the mountains, making sentences from the two tables below:

> *Example:* Everest is nearly nine times as high as Ben Nevis.
> Aconcagua is considerably higher than Mont Blanc.

nearly approximately	the same height as ... twice as high as ... x times as high as ...	
much considerably slightly	higher lower	than ...

5. Now read these statements and then complete the bar-graph, which shows the lengths of rivers.

> The Nile is approximately 16 times as long as the Thames.
> The Amazon is slightly shorter than the Nile.
> The Amazon is approximately 1,000 km longer than the Yangtse.
> The Nile is approximately twice as long as the Volga and the Parana.
> The Parana is slightly longer than the Volga.
> The Volga is considerably longer than the Euphrates.
> The Euphrates is much longer than the Thames.

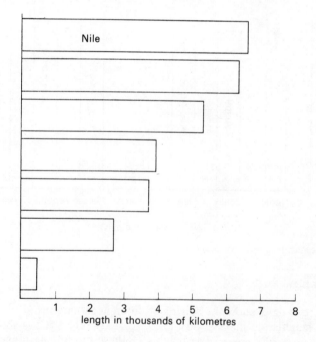

Section 2   Percentages and ratios

6. Look and read:

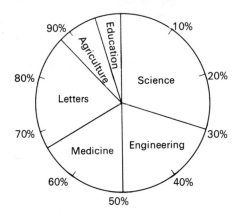

This chart shows the relative numbers of students in different faculties of a university. You can see that *the majority* (the greater part) of students study scientific or technical subjects, whereas students of letters are *in the minority*.

The proportions are approximate. They can be expressed as *percentages*. Thus, science students *constitute* approximately 30% (thirty per cent) of all students.
(Note: A consists of B = B constitutes A.)

Now complete these statements:

a) Engineering students . . . of all students.
b) 50% of all students study _____ or _____.
c) The _____ of students in the faculties of engineering and letters are approximately the same.
d) There are _____ few students of education.
e) _____ the percentage of science students, the percentage of agriculture students is relatively small.
f) In the faculty of science, 70% of the students are men and 30% are women; that is, the _____ are men and women are in the _____.
g) Approximately 15% of all students study _____.

7. Make questions and answers like the example.
The proportions can also be expressed as *ratios.*

Example: What is the ratio between students of science and students of engineering?
The *ratio between* students of science and students of engineering is 3:2. (three *to* two)

or: The ratio *of* students of science *to* students of engineering is 3:2.

Now write some similar statements about students in your college or university.

8. Look and read:

Alloys are mixtures of metals in different proportions. For example, brass is composed of *7 parts of* copper (Cu) *to 3 parts of* zinc (Zn).

Answer these:

- a) What are the percentages of copper and zinc in brass?
- b) What are A and B in the chart?
- c) What is the ratio of copper to zinc?

Now look at the charts and complete the statements:

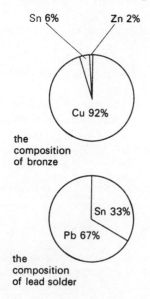

d) Bronze is made up of 1 part of _____ to 3 parts of _____ to 46 parts of _____.
e) Expressed as a _____, the composition is: copper _____%, tin _____, zinc _____.
f) The ratio between copper and the other metals is _____.

g) Lead solder consists of _____ of tin to _____ of lead.
h) Lead and tin are in a ratio of _____.

Now draw a chart to show the composition of cobalt steel:
35% cobalt (Co), 65% iron (Fe)

Section 3 Direct and inverse proportion

9. Look and read:

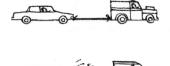

There is a *relationship* between the cross-sectional area of a rope and its strength. Its strength is *directly proportional to* its cross-sectional area.

There is also a relationship between the length of these pieces of wood and their strength. Their strength is *inversely proportional to* their length.

Do you think there is a relationship between these? If so, what is it?

> money and happiness
> the growth of a plant and sunlight
> energy and work
> the size of a person's head and his intelligence
> age and beauty

10. Look and read:

$p \propto \dfrac{1}{V}$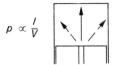

great volume/low pressure   small volume/high pressure

Pressure is inversely proportional to volume; ie *the greater* the volume, *the lower* the pressure. *Conversely*, the smaller the volume, the higher the pressure.

83

Now complete this:

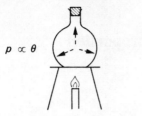

$p \propto \theta$

low temperature/low pressure   high temperature/high pressure

There is a relationship between the temperature and pressure of gases. Pressure is _____ proportional to temperature; _____ lower the temperature, ... the pressure. _____, the _____ the temperature, ... the pressure.

Make a similar statement about volume and temperature:

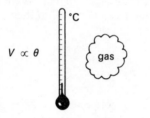
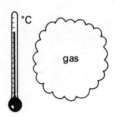

$V \propto \theta$

low temperature/small volume   high temperature/great volume

11. Look at the diagrams and say whether these statements are true or false. Correct the false statements.

$a \propto F$

a) Acceleration is directly proportional to force.
b) The greater the power of a plane's engines, the slower its acceleration.

84

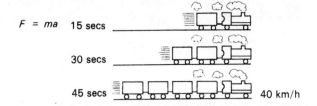

F = ma   15 secs

30 secs

45 secs .... 40 km/h

c) Acceleration is inversely proportional to mass for a given force.
d) The relationship between mass and acceleration is the same kind as that between force and acceleration.
e) The smaller the mass, the faster the acceleration.

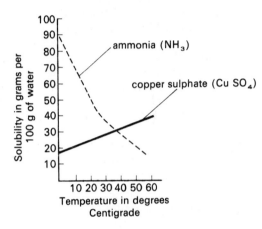

f) The solubility of copper sulphate and ammonia is inversely proportional to the temperature of the water.
g) The warmer the water, the greater the solubility of ammonia.
h) The colder the water, the less the solubility of copper sulphate.
i) The relationship between temperature and solubility is not the same for ammonia and copper sulphate.

Unit C   Revision

1. Read the following sentences and fill in the table which follows. The first one has been done for you.

a) When heated, trilead tetroxide produces lead monoxide and oxygen. It changes from an orange-red powder into a yellow solid and a colourless gas.
b) When heated, lead nitrate produces lead monoxide, nitrogen dioxide and oxygen. It is converted from white crystals into a yellow solid, a brown gas and a colourless gas.
c) Heat causes basic copper carbonate to change from a green powder to a black solid, a colourless gas and a colourless liquid. The copper carbonate produces copper oxide, carbon dioxide and water.
d) Heating potassium permanganate produces a change from violet or purple crystals to a blackish or dark-green solid. It releases oxygen and produces potassium manganate and manganese dioxide.
e) If copper sulphate crystals are heated they will change from blue crystals to a white solid and a colourless liquid. Anhydrous copper sulphate and water are produced.

Substance heated	Appearance of substance	Product	Appearance of product
Trilead tetroxide	Orange-red powder	Lead monoxide oxygen	Yellow solid and colourless gas

2. Look at this table:

Abundance of the most common elements by mass					
Crust element	%	Sea water element	%	Whole Earth element	%
Oxygen	49·4	Oxygen	91	Iron	40–50
Silicon	25·8	Hydrogen	5·7	Oxygen	22–28
Aluminium	7·5	Chlorine	2	Silicon	11–15
Iron	4·7	Sodium	1	Magnesium	9
Calcium	3·4	Magnesium	0·1	Nickel	3–6
Sodium	2·6	Sulphur	0·08	Calcium	1–2
Potassium	2·4	Calcium	0·04	Aluminium	1–2
Magnesium	2	Potassium	0·04		
Hydrogen	0·9	Bromine	0·01		
Titanium	0·5	Carbon	0·003		

Now answer these questions:

a) Which elements constitute approximately 80% of the Earth's crust?
b) What percentage of sea water do oxygen, hydrogen, and chlorine constitute?
c) Express the amounts of oxygen in the Earth's crust, in sea water, and the whole Earth as a ratio.
d) Which element has a ratio 90:20:1 in the whole Earth, in the crust, and in sea water?
e) Which element has an approximate ratio of 5:2 in the crust and in sea water?
f) Which element has a ratio of 85:1 in the crust and sea water?
g) In sea water compare the amount of chlorine with (i) oxygen (ii) bromine.
h) Compare the amount of iron in the whole Earth with the amount of iron in the crust.

3. Look at the table again and read the following passage:

Oxygen, silicon, aluminium and hydrogen together constitute approximately 80% of the Earth's crust, sea and atmosphere. Nitrogen is the main gas in the air, but is not one of the most common elements. Nitrogen forms only a small percentage of the crust and oceans, and the mass of the atmosphere is negligible compared with the total mass of the Earth.

Air is a mixture of gases. Its composition varies and depends to a large extent on plants and animals which control the amounts of oxygen and carbon dioxide by photosynthesis and respiration. Air usually also contains water vapour and dust.

If the dust is removed, the approximate composition by volume is shown in the following table:

Nitrogen	78%
Oxygen	21%
Inert gases (mostly argon)	0·93%
Carbon dioxide	0·03%

+ small quantities of other gases

Oxygen combines with metals to form oxides. In this way oxygen can be removed from a sample of air and the amount present in the sample can be measured.

Now say whether these statements are true or false. Correct the false statements:

a) Nitrogen is one of the most common elements in the earth.
b) The mass of the atmosphere is small compared with the mass of the crust.
c) The composition of the atmosphere is constant.
d) Air normally contains only gases.
e) The inert gases constitute approximately 1% of the atmosphere.
f) The inert gases include oxygen.

Unit 10 Measurement 4 Frequency, Tendency, Probability

Section 1 Frequency

1. Look at these diagrams:

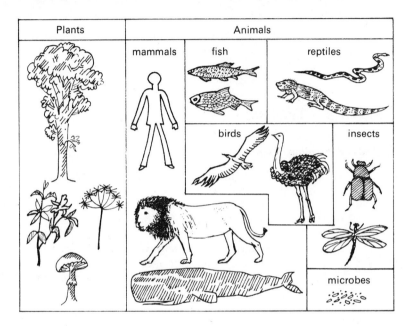

Change the statements on the left, which say how many members of a particular class possess a certain property, into statements of *frequency*, which say *how often* the property occurs:

      Example: All living things consist of cells.   *(always)*
                     Living things always consist of cells.

a) Most plants are green.   *(usually)*
b) Many birds live in trees.   *(often/frequently)*
c) Some mammals live in water.   *(sometimes)*
d) A few plants flower at night.   *(occasionally)*
e) Few fish leave the water.   *(rarely)*
f) No living things are two-dimensional.   *(never)*

2. Here are some more properties. Make statements about how often they occur:

      Examples: Plants never swim.
                       Animals usually possess tails.
                       Reptiles are always cold-blooded.

89

a) are invisible
b) breathe
c) have roots
d) lay eggs
e) suckle young
f) eat flesh
g) sing
h) climb trees
i) are warm-blooded
j) are cold-blooded

k) are three-dimensional
l) are covered with skin
m) are supported by legs
n) possess hair
o) possess lungs
p) have wings
q) are capable of flying
r) are able to swim
s) have the ability to talk

Section 2 Tendency

3. **Answer these questions:**

   a) What is the difference between these two generalisations?
      People breathe oxygen.
      People live in houses.
   b) Which one says what *always* happens?
   c) What does the other one say?
   d) Many generalisations are about what usually or *generally* occurs, ie there are *exceptions*, but these are relatively rare. Can you think of any exceptions to the statement that people live in houses?

4. **Read this:**

   Generalisations which have exceptions express a *tendency*. These statements mean the same:
      Most people live in houses.
      People generally live in houses.
      People *tend to* live in houses.

Look at these statements. Add the verb *tend to* to those which express tendency. Add *always* to those which are absolutely true. Add *sometimes, rarely, never* etc. to the others:

a) Plants are green.
b) Humans are two-legged.
c) Birds migrate in groups.
d) Birds live under water.
e) Mammals lay eggs.
f) Insects are smaller than mammals.
g) Fruit is soft.
h) Flowers are blue.

Section 3 Predicting probability

5. Look and read:

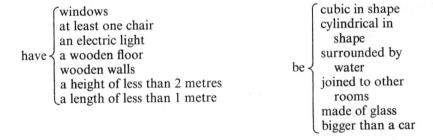

If something *always* occurs, then it *will certainly* occur.
If something *nearly always* occurs, then it *will almost certainly* occur.

		scale
usually =	will probably	100%
often =	may well	
sometimes =	may/will possibly	50%
occasionally =	might	
rarely / *seldom* =	probably will not	
never =	certainly will not	0%

The statements about frequency are based on *observation*. From them we can make *predictions* about the likelihood of something happening. Thus, we know from observation that *rooms always have walls*. Therefore we can predict that the next room we see *will certainly have walls*.

Make statements about the probability of a room having the following features:

have {
windows
at least one chair
an electric light
a wooden floor
wooden walls
a height of less than 2 metres
a length of less than 1 metre
}

be {
cubic in shape
cylindrical in shape
surrounded by water
joined to other rooms
made of glass
bigger than a car
}

6. Read this:

Some predictions depend on *conditions*.

Examples: *If the room is a laboratory*, then it probably will not contain beds.

Provided that the room is a bedroom, then it will be used for sleeping.

Make similar conditional predictions about the probability of the rooms on the left having the features on the right. (Don't use *provided that* in negative statements.)

		a shower
		beds
		a blackboard
a bedroom	contain	a cooker
a laboratory		a microscope
a workshop		an electron microscope
a bathroom	possess	a gas-supply
a hospital ward		a water-supply
a class-room		for sleeping
a kitchen		for learning
	be used	for experiments
		for washing
		for dissecting

7. Read this:

Alternative ways of predicting possibility:

	certain		100%
(extremely)	⎰ *probable* ⎱		
(fairly)	⎱ *likely* ⎰	that X *will* occur.	
It is	*possible*		50%
(fairly)	⎰ *improbable* ⎱		
(extremely)	⎱ *unlikely* ⎰		
	certain	that X *will not* occur.	0%

Now observe what percentage of students in your group have the following features:

male	right-handed	a hat
female	shoes	glasses
adult	trousers	blue eyes
child	a dress	brown eyes
animal	long hair	three legs
vegetable	short hair	a moustache
left-handed	a beard	

From these observations make predictions about the next student you meet.

Examples: It is likely that he will be adult.
If the student is female, it is probable that she will have a dress.

Section 4 Measuring probability

8. **Look and read:**

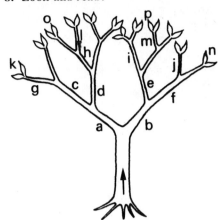

A molecule of water is moving up this tree from the roots.
Where the tree divides it is equally likely to take either branch.

Calculate the chances of the molecule reaching different points.

Examples: The *chances of* the molecule reach*ing* point *a* are *50%* or *one in two*.
The chances of it reaching point *c* are *25%* or *one in four*.

9. **Look and read:**

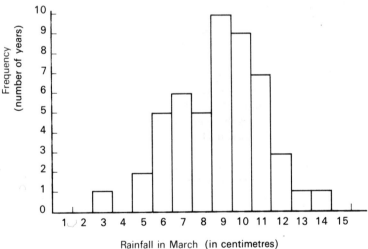

Rainfall in March (in centimetres)

This frequency diagram represents one month's average rainfall in one district over the past 50 years. It shows, for example, that the district had a rainfall of approximately 8 cm 10 times. How often did it have a rainfall of 5 cm?

93

Now look at these alternative ways of expressing probabilities:

There is a/an (extremely) (fairly) {strong, high} possibility that X will happen. — 100%

(fairly) (extremely) {weak, low, slight, remote}

no — 0%

The {possibility, probability, likelihood} that X will happen is (extremely) (fairly) {high, strong} — 100%

{low, weak, slight, remote}

nil — 0%

Make predictions like these, from the frequency diagram, about the possibility of

    a rainfall of 8 cm.
                10 cm.
                6 cm.
                4 cm.
                12 cm.
                more than 20 cm.
                less than 1 cm.
                more than 14 cm.
                between 6 and 10 cm.
                between 11 and 15 cm.

Example: The possibility that the district will have a rainfall of less than 1 cm is extremely low.

10. Look and read:

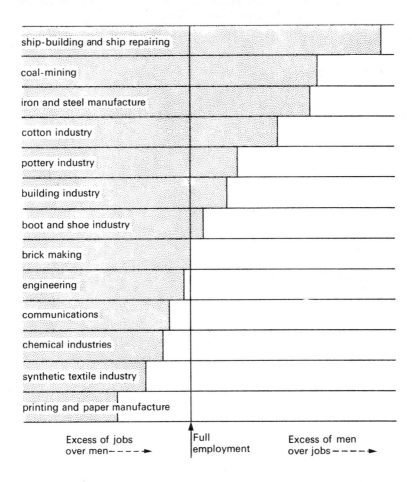

The above diagram represents the proportion of jobs to men in various industries. In ship-building, for example, there are many more workers than jobs, so there is serious unemployment.

Give the name of:
    a) an industry which has too many workers, or not enough jobs;
    b) an industry which has too few workers;
    c) an industry where the ratio of men to jobs is exactly one to one.

Unemployment is increasing. What are the relative possibilities of workers in different industries keeping or losing their jobs?

    d) If a man works in the chemical industries, is it probable that he will lose his job?
    e) In which industry is a man least likely to lose his job?

f) What possibility is there that a building worker will keep his job?
g) Is there much likelihood that a man in communications will lose his job? (much, some or little)

Now make five more predictions, using sentences like those in exercise 9.

11. Look and read:

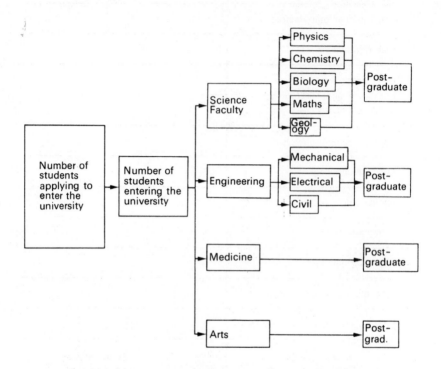

This chart shows the proportions of students entering a university and studying different subjects.

Complete these sentences:

a) (revision) The areas of the rectangles are ... to the numbers of students.
b) The next student of engineering you meet will _____ be an electrical or mechanical engineer.
c) The _____ of a student entering the university are about one in _____.
d) If a student enters the university he is more likely to study _____ than _____.
e) A student of _____ may well continue to do postgraduate studies.
f) _____ that he is accepted for the university, every student has some _____ of becoming a postgraduate.
g) A student of biology ... be in the science faculty.

h) An engineer . . . be in the arts faculty.
i) A student of _____ might become a postgraduate.
j) It is less probable that a scientist will study _____ than _____.
k) It is possible that . . .
l) There is a strong possibility that . . .
m) The probability that . . . is slight.
n) A student of science may . . .
o) Alternatively, he may . . .
p) Postgraduate students are usually . . .

Unit 11 Process 4 Method

Section 1 How things should be done

1. Look and read:

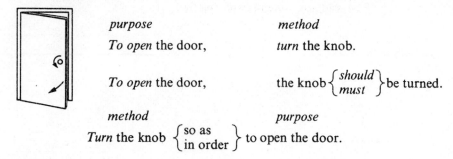

purpose	method
To open the door,	turn the knob.
To open the door,	the knob {should/must} be turned.
method Turn the knob	{so as / in order} to open the door. *purpose*

Now look at this radio. What must be done to operate it?

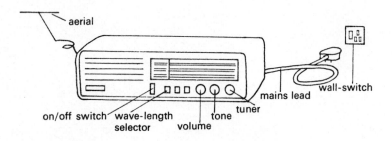

From the table below, make four sentences like each of the examples above.

purpose	method
switch on the set	press the on/off switch
adjust the volume	turn the volume knob
adjust the tone	turn the tone knob
select the right wave-length	press the wave-length selector
find the required radio station	turn the tuner
obtain better reception	fit an aerial
switch on the mains	press the wall-switch down
connect the radio with the mains	plug in the mains lead
operate the radio	

2. Look at these diagrams and instructions for operating a camera:

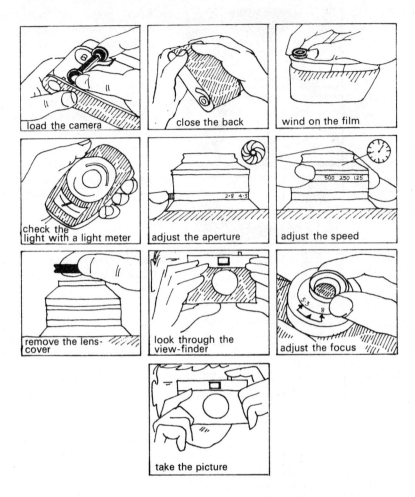

Now write ten instructions like these examples:

> *After loading* the camera, the film *should be wound on*. The camera *must be loaded before taking* the picture. *While adjusting* the focus, *look* through the view-finder.

3. Look at these examples:

> Before switching on the radio, $\begin{Bmatrix} \textit{make sure} \\ \textit{ensure} \end{Bmatrix}$ that the mains lead is plugged in.
> *Make sure* that the camera is loaded before taking the picture.

Rewrite the following precautions like the examples:

a) When you read measurements, your eye should be at right angles to the point of measurement.

b) When you read liquid measurements in cylinders, your eye should be level with the centre of the surface.

c) When empty containers are weighed, they must be dry.

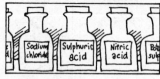

d) Before chemicals are stored, the bottles should be clearly labelled.

e) Wash your hands after you handle chemicals.

f) When acids are diluted, the acid must be added to the water.

g) Switch off the electricity before you change light bulbs.

h) When you observe the results of experiments, record the results accurately.

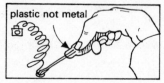

i) When electrical parts are unscrewed, the screwdriver must be insulated.

Section 2 How things may be done

4. Look again at the radio on page 98 and at these examples:

    method   *purpose*

    How $\begin{Bmatrix} may \\ can \end{Bmatrix}$ the radio *be switched on?*

    purpose   *method*

    The radio $\begin{Bmatrix} may \\ can \end{Bmatrix}$ *be switched on by pressing* the on/off switch.

Make other questions and answers like these.

5. Look at these instruments and tools and then make sentences describing what can be done with them:

Example: Temperature $\begin{Bmatrix} can \\ may \end{Bmatrix}$ be measured $\begin{Bmatrix} by\ means\ of \\ by\ using \\ with \end{Bmatrix}$ a thermometer.

purpose	*instrument or tool*
for measuring temperature	a thermometer
for viewing small objects	a microscope
for viewing distant objects	a telescope
for pouring liquids	a funnel
for heating chemicals	a bunsen burner
for weighing substances	a balance
for hammering nails	a hammer
for turning nuts & bolts	a spanner
for turning screws	a screwdriver
for measuring pressure	a barometer
for lifting things	pulleys
for testing acids and alkalis	litmus paper

6. Read this:

Alternative ways of making a magnet

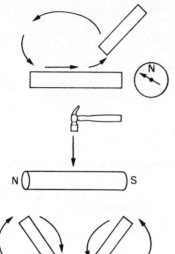

Method 1: stroke a piece of iron with a magnet. Always stroke it in the same direction. Test the new magnet with a compass.

Method 2: place a bar of iron in a north-south direction. Hit it with a hammer. Test with a compass.

Method 3: stroke a piece of iron with two magnets. Take care to use two different poles and stroke in opposite directions. Test with a compass.

Now complete these descriptions:

Method 1: A magnet may be made by . . .
The iron should always be . . .
The magnet can be . . .

Method 2: Alternatively, a bar of iron . . .
and . . . . The magnet . . .

Method 3: Another method of making a magnet is by . . .
Two different poles must . . .
and the bar must . . .

Now write descriptions with the aid of these words and diagrams:

Alternative ways of demagnetising a magnet

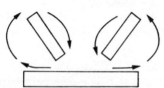

Method 1: A magnet may be demagnetised . . . placing . . . E   W hitting . . .

Method 2: Alternatively, . . . drop . . . several times . . .

Method 3: Another method of . . . heat . . . bunsen burner.

7. Look at these diagrams:

Different ways of separating materials

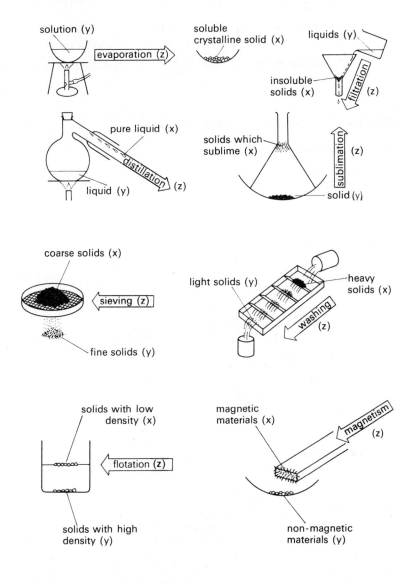

Note: x = the material to be separated
y = the other material
z = the method of separating them

Now make sequences of sentences like this example:

> One method of separating materials is by filtration.
> Alternatively, materials may be separated by sieving.
> Another method of separating materials is by magnetism.

Now write eight sentences like this example:

> A soluble crystalline solid $\begin{Bmatrix} can \\ may \end{Bmatrix}$ be separated from a solution by evaporation.

Section 3 Simple experiments

8. Look and read:

Making crystals:

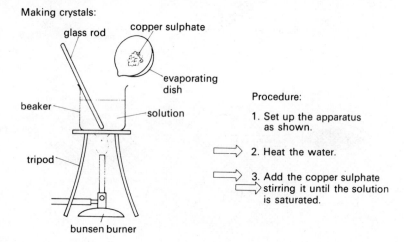

Procedure:

1. Set up the apparatus as shown.
2. Heat the water.
3. Add the copper sulphate stirring it until the solution is saturated.

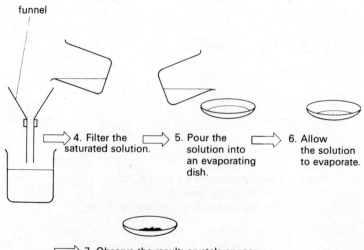

4. Filter the saturated solution.
5. Pour the solution into an evaporating dish.
6. Allow the solution to evaporate.

7. Observe the result: crystals appear.

Now complete this description of the experiment:

    The *purpose* of the experiment is to make _____.
    The *apparatus* consists of . . .
    The *substance* to be crystallised is _____.
    The *method* of carrying out the experiment is by _____.
    The *procedure* is as follows:
    First, the apparatus is set up _____.
    Then, the water is _____.
    Meanwhile, the copper sulphate _____ and stirred until . . . .
    _____, the saturated solution _____.
    Next, the solution . . . .
    _____, the solution . . . .
    Finally, the result is observed.
    The *result* is that crystals appear.

9. **Look and read:**

Separation of gunpowder:

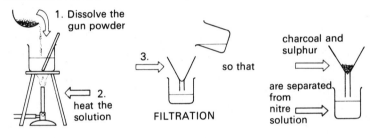

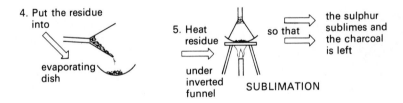

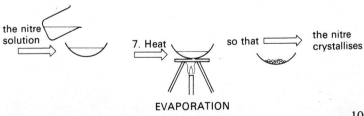

105

Now complete this description:

> The purpose . . . .
> The apparatus . . . .
> The substance . . . .
> The method . . . .
> The procedure . . . .
> The result . . . .

Unit 12 Consolidation

Section 1 Revision and reading practice

1. Look at this diagram and then read the passage below:

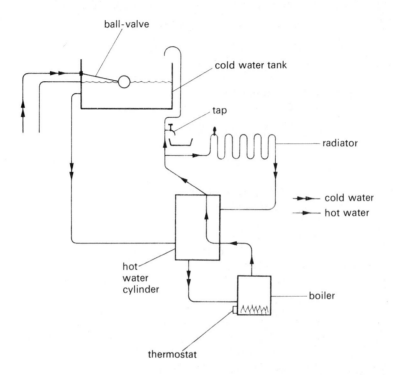

(A) A hot water system consists of a boiler and tanks for storing water. Other parts include taps and a thermostat, which is fitted on the boiler. (B) The system may also include radiators. (C) The boiler is situated at the bottom of the system. (D) It serves to heat the water. (E) Heating the water causes it (F) to rise. (G) When the hot water tap is turned on, water comes from the top of the hot water cylinder. Simultaneously, cold water flows into the cylinder from the cold water tank.

(H) The flow of water into the cold water tank is controlled by means of a ball valve. (I) The valve is connected by a bar to (J) a ball-shaped metal float. This floats because (K) it is hollow. As the tank is filled with water, the ball rises and the valve closes.

The cold water tank is situated above the hot water cylinder. Cold water flows out of the tank as a result of gravity.

The function of the thermostat is to control the temperature of the water. This ensures that the water is (L) warm enough, but prevents it from becoming (M) too hot. (N) The water has a temperature of approximately 60°C. A tank with a capacity of 60 litres is generally sufficient for an average family.

A radiator is usually rectangular in cross-section, with wide, flat sides. (O) This shape gives it a large surface area in proportion to its volume. Consequently, it gives out more heat.

The above description includes expressions which have been in previous units. The letters show their positions. Match the letters with the following:

Example: (A) is a description of structure.

a) Another description of structure.
b) A sequence of actions.
c) A statement of possibility.
d) A sufficient quantity.
e) A cause.
f) A result.
g) A simple measurement.
h) A relative measurement.
i) An excessive quantity.
j) A location.
k) A description of method.
l) A statement of function.
m) A shape.
n) Another property.

2. Now answer these questions:
a) Where is the hot water cylinder situated in relation to the cold water tank?
b) Why is part of the valve described as 'ball-shaped'?
c) Describe some of the other properties of the ball-float.
d) What causes the water to flow out of the cold water tank?
e) What is the function of the boiler?
f) What does the ball-valve consist of?
g) How much water does a hot water cylinder usually contain? (approximately)
h) What may happen if the water becomes too hot?
i) What happens as the ball-float rises to the top of the tank?
j) What is the result of heating the water?
k) How can hot water be drawn from the top of the cylinder?
l) Why does a radiator give out a large quantity of heat?
m) Why does the hot water cylinder give out less heat?

3. Read this:

What happens during an earthquake?
During an earthquake, the surface of the Earth moves. The shock is produced by waves which travel through the rock. These waves are usually the result of the movement of large masses of rock below

the surface of the Earth. Many earthquakes begin under the sea. These cause very big waves in the sea.

Now complete this:

Earthquakes are movements in the Earth's _____. _____ earthquakes, rocks move, thereby .... The waves travel to the surface and bring about the _____. Big waves in the sea result from .... During the earthquake, first _____ move, then ..., and finally ....

Now read this:

Earthquakes may take place anywhere on the Earth's surface. However, they are most likely to occur in certain regions. These are shown on the map below. Earthquake regions are usually near mountains or volcanoes. Outside these areas, earthquakes are generally weak.

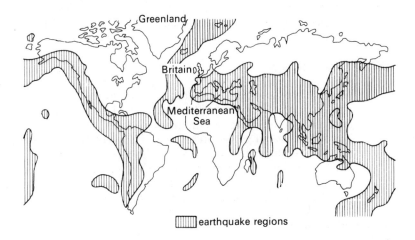

Say whether these statements are true or false. Correct the false statements.

a) Earthquakes tend to occur under the sea.
b) Earthquakes rarely take place in the north of Russia.
c) It is possible that an earthquake will take place in the north of Russia next year.
d) The chances of an earthquake occurring next month are greater in the west of South America than in the east of South America.
e) The Mediterranean Sea is surrounded by earthquake regions.
f) Earthquakes never happen in Greenland.
g) Countries where earthquakes are likely to occur include Britain.
h) In the majority of African countries there is only a slight possibility that an earthquake will occur next month.
i) The possibility of an earthquake occurring in one region may be observed by predicting the frequency of earthquakes in that region.

4. **Now read this:**

   The measurement of earthquakes
   During an earthquake the pressure waves may travel at a maximum speed of 650 km/s. The duration of earthquakes varies: an earthquake may have a duration of a second or it may continue intermittently for days. The force of an earthquake can be measured by means of special instruments. There is also a scale of measurement based on the effects of earthquakes. This scale ranges from earthquakes which are too weak to be observed by man to those which are capable of destroying everything made by man.

Answer these questions:

   a) Name four things about earthquakes which can be measured.
   b) On the scale of earthquake measurement, what is the strength of an earthquake proportional to?
   c) Why can we not observe some earthquakes?
   d) What is the effect of extremely bad earthquakes?
   e) The study of earthquakes is called 'seismology'. What is a seismometer?

5. **Finally, read this:**

   Some of the bad effects of earthquakes in towns can be prevented by making special buildings. These have two kinds of structure. In one kind of building the parts are made of light, flexible materials. The parts are woven together, like a basket. This structure and the properties of the materials enable the building to move without breaking. The other kind of building is like a box in structure. It is made of heavy, rigid materials. The lower part of the building must have a much greater mass than the upper part.

Now complete these:

This is a basket. It is strong because .... and because ....

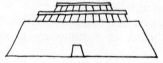

This building is strong because... and because ....

This building is not good for earthquake regions because the roof is . . . .

Section 2   Appearance, fact and proof

6. Look and read:

Is this picture two-dimensional or three-dimensional?

It *appears to be* three-dimensional, but *in fact* it is two-dimensional.

Now complete these:

a)

This . . . a hemisphere, . . . consists of curved and straight lines.

b)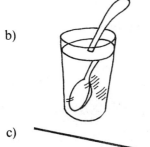

A spoon in a glass of water . . . broken, . . . not broken.

c)

These lines *converge*.

These lines are _____.

If you look along a road, the sides of the road appear to _____, but in fact . . . .

d)

_____ the Sun approaches the horizon, its diameter _____ become bigger. The farther it is from the _____, the smaller it appears. The nearer it is to the horizon, ... Its diameter appears to be . . . to its distance from the horizon.

_____, the size of the Sun does not vary; it is _____. The *apparent* change in size is caused by dust in the air near the horizon.

e) 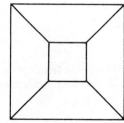 Look at this figure. It appears to be _____-dimensional, but in fact . . . . The small square sometimes appears to be behind the big square. Sometimes it appears to be _____ the big square. In fact it is _____ the big square.

7. Now look at these figures and answer the questions, saying what they appear to be and what they are in fact:

a) 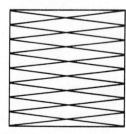 Is this a rectangle or a square?

b) 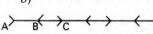 Have lines AB and BC the same length or different lengths?

112

c)

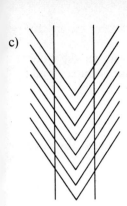

Do the vertical lines converge or are they parallel?

d)

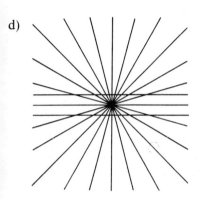

Are the top and bottom horizontal lines curved or straight?

e)

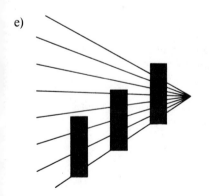

Is the right-hand rectangle higher than the left-hand rectangle or are their heights the same?

8. Now read these sentences:

We *can prove that* figure (a) is a square by measuring the sides. *To show that* the vertical lines in figure (c) are parallel, a ruler should be placed along them.

Complete these:

a) We can _____ that lines AB and BC in figure (b) have the same length by . . . .
b) . . . the horizontal lines in figure (d) are in fact straight, ...
c) To prove that the heights of the rectangles in figure (e) . . . , they . . . .
d) By placing a ruler along the vertical lines in figure (c), we can . . . .

Section 3 More simple experiments

9. Read these instructions:

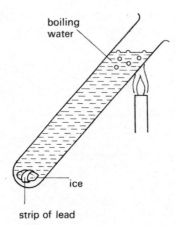

strip of lead

Purpose:
To show that water is a bad conductor of heat.

Procedure:
Wrap a strip of lead round a piece of ice. Drop the ice into a test-tube which contains water.

Heat the water at the top of the test-tube until it boils.

Observation: When the water boils, the ice does not melt.
Conclusion: Water is a bad conductor of heat.

Now complete this:

a) By doing this experiment, we can prove that . . . .
b) The apparatus for this experiment consists of . . . .
c) After the ice is dropped into the test-tube . . . .
d) The ice sinks to the bottom of the test-tube because . . . .
e) During this experiment, we observe that . . . .
f) The conclusion drawn from this experiment is that . . . .

10. Now read this description:

Different materials conduct heat at different speeds. This can be shown by doing a simple experiment. Some rods are covered with wax and attached to a metal tank. The rods have the same size but they are made of different materials. The tank is filled with water.

The wax on the rods soon starts to melt. On the copper rod, a large amount of wax melts. Hardly any wax melts on the wooden rod. On the other rods, the amount of melted wax varies. This shows that some materials are good conductors of heat, while others are bad conductors of heat.

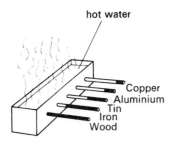

Answer these questions:

a) What is the purpose of this experiment?
b) What does the apparatus consist of?
c) What is the procedure?
d) What can we observe during the experiment?
e) What can we conclude from the experiment?
f) What property of copper does this experiment show?
g) Why do cooking pots often have wooden handles?

11. Look at this diagram:

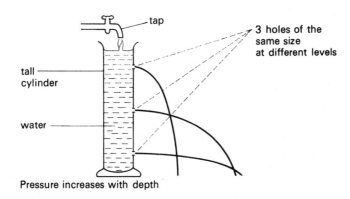

Now complete these sentences:

    Purpose: To show that . . .

    Apparatus: This consists of . . .

    Procedure: Place the cylinder . . .
                     Turn on the tap so that . . .

    Observation: The water comes most quickly . . . and most slowly . . .

    Conclusion: Pressure is higher at . . . than near . . . of the water. Therefore, pressure increases . . . ie pressure is proportional . . .

Glossary

This list gives the pronunciations of the technical and less common words used in this book and definitions of those words that are not fully explained in the text or diagrams. An asterisk (*) means that a word in the definition is itself explained in the Glossary. The number after each entry indicates the unit in which the word first appears.

Pronunciations are shown in the system that is used in the new Longman *Dictionary of Contemporary English*. The symbols are shown in this table, with a key word for each. The letters printed in **bold type** represent the sound value of the symbol.

Consonants

p	pea	f	few	ʃ	fishing	h	hot
b	bay	v	view	ʒ	pleasure	m	sum
t	tea	θ	thing	tʃ	choose	n	sun
d	day	ð	then	dʒ	jump	ŋ	sung
k	key	s	soon	l	led	j	yet
g	gay	z	zoo	r	red	w	wet

Vowels

iː	sheep	ɔː	caught	eɪ	make	ɪə	here
ɪ	ship	ʊ	put	əʊ	note	eə	there
e	bed	uː	boot	aɪ	bite	ʊə	poor
æ	bad	ʌ	cut	aʊ	now	eɪə	player
ɑː	calm	ɜː	bird	ɔɪ	boy	əʊə	lower
ɒ	cot	ə	about			aɪə	tire
						aʊə	tower
						ɔɪə	employer

Notes

1. A small raised /ʳ/ at the end of a word means that the /r/ is pronounced if a vowel follows (at the beginning of the text word), but not otherwise. For example, *far* /fɑːʳ/ means that *far away* is pronounced /fɑːr əweɪ/ but *far down* is /fɑː daʊn/.

2. The italic /ə/ means that the sound /ə/ can be used but is often omitted. It may be found before the consonants /m, n, ŋ, l, r/ in certain positions. For example, *travel* /ˈtrævəl/ means that the pronounciation /ˈtrævəl/ is possible but /ˈtrævl/ may be more common.

3. The mark /ˈ/ means that the following syllable has *main stress*, and /ˌ/ means that the following syllable has *secondary stress*. For example, *understand* /ˌʌndəˈstænd/.

ability /əˈbɪlətɪ/ being able to do something 5
absorbent /əbˈzɔːbənt/ able to take in liquid 3
absorption /əbˈzɔːpʃn/ 6
accelerator /əkˈseləreɪtəʳ/ 5
Aconcagua /ˌækɒŋˈkɑːgwə/ 4
adequate /ˈædɪkwət/ enough 7
adjacent /əˈdʒeɪsənt/ 2
adjust /əˈdʒʌst/ make a thing right 5
aerial /ˈeərɪəl/ 1
alcohol /ˈælkəhɒl/ 4
alkali /ˈælkəlaɪ/ 8
alloy /ˈælɔɪ/ 9
aluminium /ˌæljʊˈmɪnjəm/ 7
Amazon /ˈæməzən/ 4
ammonia /əˈməʊnjə/ 9
amoeba /əˈmiːbə/ 4
ampere /ˈæmpeəʳ/ 4
anaemia /əˈniːmjə/ 7
aneroid barometer /ˈænərɔɪd bəˈrɒmɪtəʳ/ barometer worked by a vacuum box with springy walls B
aorta /eɪˈɔːtə/ 6
apparatus /ˌæpəˈreɪtəs/ 1
appearance/ əˈpɪərəns/ way something looks 12
approximately /əˈprɒksɪmətlɪ/ 4
April /ˈeɪprəl/ 4
area /ˈeərɪə/ 1
argon /ˈɑːgɒn/ (A) 7
atmosphere /ˈætməsfɪəʳ/ 3
atmospheric /ˌætməsˈferɪk/ 5
attract /əˈtrækt/ pull towards 8
August /ˈɔːgəst/ 4
auricle /ˌɔːrɪkl/ 8
average /ˈævərɪdʒ/ 4
axle /ˈæksəl/ 3

bacteria /bækˈtɪərɪə/ 4
balloon /bəˈluːn/ bag filled with gas lighter than air 8
barometer /bəˈrɒmɪtəʳ/ 5
bromine /ˈbrəʊmiːn/ (Br) C
bunsen burner /ˌbʌnsən ˈbɜːnəʳ/ 2
burette /bjʊəˈret/ 7

cadmium /ˈkædmɪəm/ (Cd) 2
calcium /ˈkælsɪəm/ (Ca) 7
capacity /kəˈpæsətɪ/ (1) amount a container can hold 4; (2) being able to do something 5
carbohydrate /ˌkɑːbəʊˈhaɪdreɪt/ 7
carbon /ˈkɑːbən/ (C) 7
carbonic /kɑːˈbɒnɪk/ 8
casing /ˈkeɪsɪŋ/ outer covering 3

centi- /ˈsentɪ/ hundredth 4
charcoal /ˈtʃɑːkəʊl/ black fuel made by burning wood slowly with little air 8
chloride /ˈklɔːraɪd/ 7
chlorine /ˈklɔːriːn/ (Cl) 8
chlorophyll /ˈklɒrəfɪl/ 8
chromium /ˈkrəʊmɪəm/ (Cr) 2
circular /ˈsɜːkjələʳ/ 1
circumference /səˈkʌmfərəns/ 4
cobalt /ˌkəʊbɔːlt/ (Co) 2
column /ˈkɒləm/ 2
combustible /kəmˈbʌstəbl/ 1
combustion /kəmˈbʌstʃən/ 8
compare /kəmˈpeəʳ/ look at things to see how they are the same or different 9
compass /ˈkʌmpəs/ 11
compound /ˈkɒmpaʊnd/ substance* formed by uniting different elements* 3
compress /kəmˈpres/ press together into a smaller space 6
concave /ˈkɒŋkeɪv/ 2
conclude /kənˈkluːd/ decide as a result of reasoning 12
conical /ˈkɒnɪkl/ 1
considerable /kənˈsɪdərəbl/ 7
constant /ˈkɒnstənt/ unchanging 4
constitute /ˈkɒnstɪtjuːt/ make up C
consume /kənˈsjuːm/ eat, use up 7
conversely /kənˈvɜːslɪ/ 9
convert /kənˈvɜːt/ change 8
crystal /ˈkrɪstl/ 7
crystalline /ˈkrɪstəlaɪn/ having a regular arrangement of atoms 11
crystallise /ˈkrɪstəlaɪz/ form crystals 8
cubic /ˈkjuːbɪk/ 1
cycle /ˈsaɪkl/ repeating set of events 6
cylinder /ˈsɪlɪndəʳ/ 1
cylindrical /sɪˈlɪndrɪkl/ 1
cytoplasm /ˈsaɪtəʊplæzəm/ 3

decay /dɪˈkeɪ/ decompose* and rot as a result of bacterial action 6
December /dɪˈsembəʳ/ 4
deci- /ˈdesɪ/ tenth 4
decompose /ˌdiːkəmˈpəʊz/ break down into simpler substances* 6
degree /dɪˈgriː/ 4
density /ˈdensətɪ/ 4
depend on /dɪˈpend ˌɒn/ be controlled and influenced by 8
desert /ˈdezət/ 2
device /dɪˈvaɪs/ instrument* 5
diagonal /daɪˈægənəl/ 1
diameter /daɪˈæmɪtəʳ/ 4

diamond /ˈdaɪəmənd/ very hard brilliant precious stone 1
dilute /daɪˈluːt, daɪˈljuːt/ make weaker or thinner 11
dimension /daɪˈmenʃn/ measurement, eg length, width, height 1
dimensional /daɪˈmenʃənəl/ 1
disintegrate /dɪˈsɪntəgreɪt/ break into small pieces 8
dissect /dɪˈsekt/ cut up in order to examine 10
distil /dɪˈstɪl/ make a liquid into vapour, then make the vapour into liquid 5
distributed /dɪˈstrɪbjuːtɪd/ found in different places 2
drains /dreɪnz/ pipes or channels to carry away waste 6
duration /dʒʊəˈreɪʃn, djʊəˈreɪʃn/ time during which something lasts or exists 4
dynamo /ˈdaɪnəməʊ/ 5

effect /ɪˈfekt/ result of a cause 8
eg /ˌiːˈdʒiː/ = for example 1
electrolyte /ɪˈlektrəlaɪt/ 3
element /ˈelɪmənt/ substance* which cannot be split up into simpler substances* 2
emerge /ɪˈmɜːdʒ/ come out 6
engineering /ˌendʒɪˈnɪərɪŋ/ 9
etiolated /ˈiːtɪəʊleɪtɪd/ (of plants) pale owing to lack of light 8
exactly /ɪgˈzæktlɪ/ 4
except /ɪkˈsept/ but not including 1
exception /ɪkˈsepʃn/ something not covered by a rule 10
excess /ɪkˈses/ 7
excessive /ɪkˈsesɪv/ 7
exposed /ɪkˈspəʊzd/ laid open 8
exterior ɪkˈstɪərɪə/ 2
external /ɪkˈstɜːnəl/ of the outside 5
extinguish /ɪkˈstɪŋgwɪʃ/ 6
extremes /ɪkˈstriːmz/ highest and lowest points 4

fact /fækt/ real truth 12
February /ˈfebrʊərɪ/ 4
fertile /ˈfɜːtaɪl/ (of land) producing many crops 7
fertilise /ˈfɜːtɪlaɪz/ make fertile* or productive 6
fluid /ˈfluːɪd/ substance* which flows—liquid or gas 1
formula /ˈfɔːmjələ/ pl. **formulae** /ˈfɔːmjʊliː/ 4

friction /ˈfrɪkʃn/ force which tries to stop one thing from slipping over another 1
fulcrum /ˈfʌlkrəm/ point on which a lever moves A
function /ˈfʌŋkʃn/ special purpose or use of something 5

galaxy /ˈgæləksɪ/ group of stars 4
gaseous /ˈgeɪsɪəs, ˈgæsɪəs/ 1
gauge /geɪdʒ/ 5
generalisation /ˌdʒenərəlaɪˈzeɪʃn/ statement about things in general 10
generally /ˈdʒenərəlɪ/ usually, mostly 1
germination /ˌdʒɜːmɪˈneɪʃn/ 6
give off /ˌgɪv ˈɒf/ emit, send off 8
glycogen /ˈglaɪkəʊdʒen/ 5
green wood /ˈgriːn ˌwʊd/ live wood 1

hafnium /ˈhæfnjəm/ (Hf) 2
height /haɪt/ 4
helium /ˈhiːljəm/ (He) 7
hollow /ˈhɒləʊ/ having an empty space inside 2
horizon /həˈraɪzn/ 6
horizontal /ˌhɒrɪˈzɒntl/ 1
hydrochloric /ˌhaɪdrəˈklɒrɪk/ 3
hydrogen /ˈhaɪdrədʒən/ (H) 3

identify /aɪˈdentɪfaɪ/ say what something is, give it a name 8
ie /ˌaɪˈiː/ = that is 4
ignite /ɪgˈnaɪt/ 8
immediately /ɪˈmiːdɪətlɪ/ at once without a pause 6
inanimate /ɪnˈænɪmət/ without life 5
inert /ɪˈnɜːt/ (of gases) forming no chemical compounds* C
instal /ɪnˈstɔːl/ fix in position for use 6
instrument /ˈɪnstrəmənt/ apparatus used for precise work 5
interior /ɪnˈtɪərɪəʳ/ 2
intermittently /ˌɪntəˈmɪtəntlɪ/ at intervals 12
internal /ɪnˈtɜːnəl/ of the inside 5
inverse /ˈɪnvɜːs/ 9
iodine /ˈaɪədiːn/ (I) 8
iridium /aɪˈrɪdjəm/ (Ir) 2
iron ˈaɪən/ (Fe) 2
irrigate /ˈɪrɪgeɪt/ supply water to land by means of canals etc. 7

January /ˈdʒænjʊərɪ/ 4
July /dʒuːˈlaɪ/ 4
June /dʒuːn/ 4

Kilimanjaro /ˌkɪlɪmənˈdʒɑːrəʊ/ 4
kilo- /ˈkɪlə/ thousand/ 4
kilometre /kɪˈlɒmɪtəʳ, ˈkɪləˌmiːtəʳ/ 4
krypton /ˈkrɪptɒn/ (Kr) 7

label /ˈleɪbl/ put a name on something 11
laboratory /ləˈbɒrətrɪ/ 10
lanthanum /ˈlænθənəm/ (La) 2
lay eggs /ˌleɪ ˈegz/ produce eggs 10
lay foundations /ˌleɪ faunˈdeɪʃnz/ build the supporting base of a building 6

lead /led/ (Pb) 7
length /leŋθ/ 4
level /ˈlevəl/ make flat 6
liberate /ˈlɪbəreɪt/ release a gas from combination 8
likelihood /ˈlaɪklɪhʊd/ probability* 10
liquefaction /ˌlɪkwɪˈfækʃn/ making liquid 6
lithium /ˈlɪθjəm/ (Li) 7
litmus paper /ˈlɪtməs ˌpeɪpəʳ/ chemical indicator which turns red with an acid, blue with an alkali 8
litre /ˈliːtəʳ/ 4
location /ləʊˈkeɪʃn/ 2
longitudinal /ˌlɒndʒɪˈtjuːdɪnəl/ 1

magnesium /mægˈniːzɪəm/ (Mg) 7
majority /məˈdʒɒrətɪ/ greater number or amount 9
mammal /ˈmæməl/ 10
manganese /ˈmæŋgəniːz/ (Mn) 2
manufacture /ˌmænjʊˈfæktʃəʳ/ production 10
March /mɑːtʃ/ 4
mass /mæs/ quantity of material in a body 4
material /məˈtɪərɪəl/ 1
maximum /ˈmæksɪməm/ greatest size, number etc. 4
May /meɪ/ 4
measurement /ˈmeʒəmənt/ 4
mercury /ˈmɜːkjərɪ/ (Hg) 2
methane /ˈmiːθeɪn/ (CH_4) 8
method /ˈmeθəd/ way of doing something 11
micrometer /maɪˈkrɒmɪtəʳ/ 4
migrate /maɪˈgreɪt/ move from one place to another 10
milli- /ˈmɪlɪ/ thousandth 4
minimum /ˈmɪnɪməm/ smallest size, number etc. 4
minority /maɪˈnɒrətɪ/ smaller number or amount 9
minute /maɪˈnjuːt/ very small 4

molecule /ˈmɒlɪkjuːl/ 4
molybdenum /məˈlɪbdənəm/ (Mo) 2

narrow /ˈnærəʊ/ (of range) small 4
negligible /ˈneglɪdʒəbl/ 7
neon /ˈniːɒn/ (Ne) 1
neutralise /ˈnjuːtrəlaɪz/ make neutral (neither acid nor alkaline) 8
niobium /naɪˈəʊbjəm/ (Nb) 2
nitrogen /ˈnaɪtrədʒən/ (N) 7
November /nəʊˈvembəʳ/ 4

obtain /əbˈteɪn/ get 8
occur /əˈkɜːʳ/ happen, be found 6
ocean /ˈəʊʃn/ 2
October /ɒkˈtəʊbəʳ/ 4
ohm /əʊm/ 4
opaque /əʊˈpeɪk/ 1
organ /ˈɔːgən/ part of a living body with a special function 5
organic /ɔːˈgænɪk/ found in living things 3
osmium /ˈɒzmɪəm/ (Os) 2
oxide /ˈɒksaɪd/ 8
oxygen /ˈɒksɪdʒən/ (O) 1
oxygenate /ˈɒksɪdʒəneɪt/ 5

palladium /pəˈleɪdjəm/ (Pd) 2
pan /pæn/ part of a balance on which things are placed for weighing A
parallel /ˈpærəlel/ 1
per (normally unstressed) /pəʳ/ 4
percentage /pəˈsentɪdʒ/ number or rate in each hundred
phosphorous /ˈfɒsfərəs/ (P) 7
photosynthesis /ˌfəʊtəʊ ˈsɪnθəsɪs/ building up of sugars and starches in the green cells of a plant, by means of chlorophyll, in the presence of sunlight 8
pi /paɪ/ π ratio of circumference of a circle to its diameter (about 3·14159)
pipette /pɪˈpet/ 7
platinum /ˈplætɪnəm/ (Pt) 2
point of condensation /ˌpɔɪnt əv kɒndenˈseɪʃn/ point at which vapour becomes liquid 1
boiling point /ˈbɔɪlɪŋ ˌpɔɪnt/ point at which liquid becomes vapour 1
freezing point /ˈfriːzɪŋ ˌpɔɪnt/ point at which a liquid becomes solid 1
melting point /ˈmeltɪŋ ˌpɔɪnt/ point at which a solid becomes liquid 1
pollen /ˈpɒlən/ flower-dust carrying male cells 6

pollination /ˌpɒlɪˈneɪʃn/ carrying of pollen* to the female part of a flower 6
population /ˌpɒpjəˈleɪʃn/ number of people 7
over/under-populated /ˈəʊvə/ˈʌndə ˌpɒpjəleɪtɪd/ having too many/too few people 7
possess /pəˈzes/ have 7
potassium /pəˈtæsɪəm/ (K) 7
potential difference /pəˌtenʃl ˈdɪfrəns/ difference of electric pressure 4
precede /prɪˈsiːd/ be before or in front of 6
precipitation /prɪˌsɪpɪˈteɪʃn/ fall of rain, snow etc. 6
prediction /prɪˈdɪkʃn/ saying in advance what will happen 10
pressure /ˈpreʃəʳ/ B
prior to /ˈpraɪə tʊ/ before 6
prism /ˈprɪzəm/ 1
proactinium /ˌprəʊˈæktɪnɪəm/ (Pa) 8
probability /ˌprɒbəˈbɪləti/ chance 10
procedure /prəˈsiːdʒəʳ/ way of doing something 11
process /ˈprəʊses/ 5
property /ˈprɒpəti/ special quality belonging to something 1
proportion /prəˈpɔːʃn/ size of something when thought of as part of a whole, or in relation to something else 9
prove /pruːv/ show that something is true 12
provided that /prəˈvaɪdɪd ðət/ (only) if 10

quantity /ˈkwɒntəti/ 7

radiate /ˈreɪdɪeɪt/ send out energy 5
radiator /ˈreɪdɪeɪtəʳ/ 5
radioactive /ˌreɪdɪəʊˈæktɪv/ 8
radium /ˈreɪdɪəm/ (Ra) 7
radius /ˈreɪdɪəs/ 4
range /reɪndʒ/ distance or change between fixed points 4
ratio /ˈreɪʃɪəʊ/ 9
rectangular /rekˈtæŋgjələʳ/ 6
refrigerator /rɪˈfrɪdʒəˌreɪtəʳ/ 6
region /ˈriːdʒən/ large area 12
relatively /ˈrelətɪvli/ when compared* with others 9
remote /rɪˈməʊt/ (of chance) very small 10
repel /rɪˈpel/ push away (by magnetism) 8
reptile /ˈreptaɪl/ 10

residue /ˈrezɪdʒuː, ˈrezɪdjuː/ that which remains 11
respiration /ˌrespɪˈreɪʃn/ 8
result /rɪˈzʌlt/ that which a cause produces 8
rhenium /ˈriːnɪəm/ (Re) 2
rhodium /ˈrəʊdɪəm/ (Rh) 2
rigid /ˈrɪdʒɪd/ 1
roughly /ˈrʌflɪ/ 1
row /rəʊ/ 2
ruthenium /ruːˈθiːnɪəm/ (Ru) 2

saturated solution /ˌsætʃʊreɪtɪd səˈluːʃn/ a solution which contains so much of the dissolved substance* that no more can be dissolved at that temperature 11
scale /skeɪl/ (1) series of marks at regular intervals for the purpose of measuring A; (2) size of a map etc. in relation to what it represents 9
scandium /ˈskændɪəm/ (Sc) 2
seal /siːl/ close tightly B
seismology /saɪzˈmɒlədʒɪ/ 12
September /səpˈtembəʳ/ 4
sequence /ˈsiːkwəns/ order of events 6
sieve /sɪv/ press through wire net in order to separate different substances* 11
silicon /ˈsɪlɪkɒn/ (Si) 7
simultaneous /ˌsɪməlˈteɪnɪəs/ happening at the same time 6
site /saɪt/ piece of ground where a building is or will be built 6
sodium /ˈsəʊdɪəm/ (Na) 7
solder /ˈsəʊldəʳ/ alloy used to join metals together 9
soluble /ˈsɒljʊbl/ 1
spatial /ˈspeɪʃl/ in relation to space 4
speedometer /spiːˈdɒmɪtəʳ/ 5
spherical /ˈsferɪkl/ 1
spiral /ˈspaɪərəl/ A
state /steɪt/ condition 1
stir /stɜːʳ/ mix with a cylindrical rod 11
store /stɔːʳ/ keep in one place for later use 5
stroke /strəʊk/ rub gently 11
strontium /ˈstrɒntɪəm/ (Sr) 7
sublime /səˈblaɪm/ change a solid into a vapour 8
substance /ˈsʌbstəns/ (particular kind of) matter 3
suckle /ˈsʌkl/ feed with milk from the breast or udder 10
sufficient /səˈfɪʃnt/ enough 7
suitable /ˈsuːtəbl, ˈsjuːtəbl/ right or .... proper for a particular purpose 1

121

sulphur /ˈsʌlfəʳ/ (S) 7
supply /səˌplaɪ/ give or provide 5
support /səˈpɔːt/ provide enough for 7
surface /ˈsɜːfɪs/ 1
surface tension /ˌsɜːfɪsˈtenʃn/ 8
swell /swel/ grow larger 6
synthetic /sɪnˈθetɪk/ made by man, not natural 10

tantalum /ˈtæntələm/ (Ta) 2
tapering /ˈteɪpərɪŋ/ 1
technetium /tekˈniːsɪəm/ (Tc) 2
temperate /ˈtempərət/ (of climate) not too hot or cold 7
temperature ˈtempərətʃəʳ/ 4
test /test/ find the quality, composition etc. by examination 11
thermometer /θəˈmɒmɪtəʳ/ 5
thermostat /ˈθɜːməˌstæt/ instrument* which keeps temperature at one level 12
thorium /ˈθɔːrɪəm/ (Th) 8
titanium /taɪˈteɪnɪəm/ (Ti) 2
total /ˈtəʊtl/ added together 4
tough /tʌf/ 1
translucent /trænzˈluːsənt/ 1
transparent /trænsˈpærənt/ 1
triangular /traɪˈæŋgjələʳ/ 1
tripod /ˈtraɪpɒd/ 1
trough /trɒf/ 1

tungsten /ˈtʌŋstən/ (W) 2

unit /ˈjuːnɪt/ quantity or amount used as a standard measurement 4
uranium /juːˈreɪnɪəm/ (U) 7

valve /vælv/ 8
vanadium /vəˈneɪdjəm/ (V) 2
vaporised /ˈveɪpəraɪzd/ converted* into vapour 5
vary /ˈveərɪ/ be different, change 4
vena cava /ˌviːnə ˈkeɪvə/ B
ventricle /ˈventrɪkl/ 8
vertical /ˈvɜːtɪkl/ 1
volcano /vɒlˈkeɪnəʊ/ 6
volt /vəʊlt/ 4
volume /ˈvɒljuːm/ 4

watt /wɒt/ 4
wax /wæks/ soft yellow substance produced by bees, also obtained from petroleum 12
weight /weɪt/ 4
width /wɪtθ/ 4

xenon /ˈzenɒn/ (Xe) 7

yttrium /ˈɪtrɪəm/ (Y) 2

zirconium /zɜːˈkəʊnjəm/ (Zr) 2